Learn

Eureka Math®
Grade 4
Module 3

Published by Great Minds®.

Copyright © 2018 Great Minds®.

Printed in the U.S.A.

This book may be purchased from the publisher at eureka-math.org.

10 9 8 7 6 5 4 3 2 1

ISBN 978-1-64054-066-8

G4-M3-L-05.2018

Learn ◆ Practice ◆ Succeed

Eureka Math® student materials for *A Story of Units®* (K–5) are available in the *Learn, Practice, Succeed* trio. This series supports differentiation and remediation while keeping student materials organized and accessible. Educators will find that the *Learn, Practice,* and *Succeed* series also offers coherent—and therefore, more effective—resources for Response to Intervention (RTI), extra practice, and summer learning.

Learn

Eureka Math Learn serves as a student's in-class companion where they show their thinking, share what they know, and watch their knowledge build every day. *Learn* assembles the daily classwork—Application Problems, Exit Tickets, Problem Sets, templates—in an easily stored and navigated volume.

Practice

Each *Eureka Math* lesson begins with a series of energetic, joyous fluency activities, including those found in *Eureka Math Practice.* Students who are fluent in their math facts can master more material more deeply. With *Practice,* students build competence in newly acquired skills and reinforce previous learning in preparation for the next lesson.

Together, *Learn* and *Practice* provide all the print materials students will use for their core math instruction.

Succeed

Eureka Math Succeed enables students to work individually toward mastery. These additional problem sets align lesson by lesson with classroom instruction, making them ideal for use as homework or extra practice. Each problem set is accompanied by a Homework Helper, a set of worked examples that illustrate how to solve similar problems.

Teachers and tutors can use *Succeed* books from prior grade levels as curriculum-consistent tools for filling gaps in foundational knowledge. Students will thrive and progress more quickly as familiar models facilitate connections to their current grade-level content.

Students, families, and educators:

Thank you for being part of the *Eureka Math*® community, where we celebrate the joy, wonder, and thrill of mathematics.

In the *Eureka Math* classroom, new learning is activated through rich experiences and dialogue. The *Learn* book puts in each student's hands the prompts and problem sequences they need to express and consolidate their learning in class.

What is in the Learn book?

Application Problems: Problem solving in a real-world context is a daily part of *Eureka Math*. Students build confidence and perseverance as they apply their knowledge in new and varied situations. The curriculum encourages students to use the RDW process—Read the problem, Draw to make sense of the problem, and Write an equation and a solution. Teachers facilitate as students share their work and explain their solution strategies to one another.

Problem Sets: A carefully sequenced Problem Set provides an in-class opportunity for independent work, with multiple entry points for differentiation. Teachers can use the Preparation and Customization process to select "Must Do" problems for each student. Some students will complete more problems than others; what is important is that all students have a 10-minute period to immediately exercise what they've learned, with light support from their teacher.

Students bring the Problem Set with them to the culminating point of each lesson: the Student Debrief. Here, students reflect with their peers and their teacher, articulating and consolidating what they wondered, noticed, and learned that day.

Exit Tickets: Students show their teacher what they know through their work on the daily Exit Ticket. This check for understanding provides the teacher with valuable real-time evidence of the efficacy of that day's instruction, giving critical insight into where to focus next.

Templates: From time to time, the Application Problem, Problem Set, or other classroom activity requires that students have their own copy of a picture, reusable model, or data set. Each of these templates is provided with the first lesson that requires it.

Where can I learn more about Eureka Math *resources?*

The Great Minds® team is committed to supporting students, families, and educators with an ever-growing library of resources, available at eureka-math.org. The website also offers inspiring stories of success in the *Eureka Math* community. Share your insights and accomplishments with fellow users by becoming a *Eureka Math* Champion.

Best wishes for a year filled with aha moments!

Jill Diniz

Jill Diniz
Director of Mathematics
Great Minds

The Read–Draw–Write Process

The *Eureka Math* curriculum supports students as they problem-solve by using a simple, repeatable process introduced by the teacher. The Read–Draw–Write (RDW) process calls for students to

1. Read the problem.

2. Draw and label.

3. Write an equation.

4. Write a word sentence (statement).

Educators are encouraged to scaffold the process by interjecting questions such as

- What do you see?

- Can you draw something?

- What conclusions can you make from your drawing?

The more students participate in reasoning through problems with this systematic, open approach, the more they internalize the thought process and apply it instinctively for years to come.

Contents

Module 3: Multi-Digit Multiplication and Division

Name _____   Date _____

1. Determine the perimeter and area of rectangles A and B.

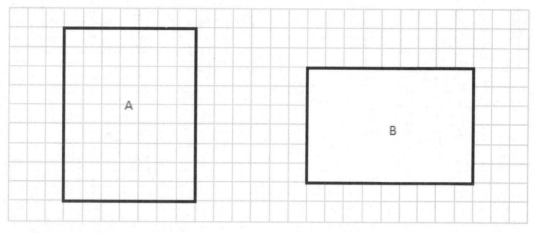

a. A = _____   A = _____

b. P = _____   P = _____

2. Determine the perimeter and area of each rectangle.

a.   b.

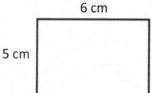

P = _____

A = _____

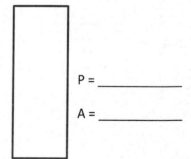

P = _____

A = _____

3. Determine the perimeter of each rectangle.

    a.

    166 m

    99 m

    P = _____

    b.

    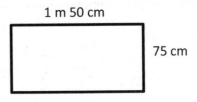

    1 m 50 cm

    75 cm

    P = _____

4. Given the rectangle's area, find the unknown side length.

    a.

    8 cm

    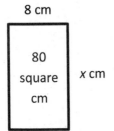

    80 square cm

    x cm

    x = _____

    b.

    7 cm

    49 square cm

    x cm

    x = _____

Lesson 1:   Investigate and use the formulas for area and perimeter of rectangles.

EUREKA MATH

5. Given the rectangle's perimeter, find the unknown side length.

   a. P = 120 cm

   20 cm

   x cm

   x = _____

   b. P = 1,000 m

   x m

   250 m

   x = _____

6. Each of the following rectangles has whole number side lengths. Given the area and perimeter, find the length and width.

   a. P = 20 cm

   l = _____

   24 square cm

   w = _____

   b. P = 28 m

   w = _____

   24 square m

   l = _____

Lesson 1:   Investigate and use the formulas for area and perimeter of rectangles.

3

EUREKA MATH

Name _____   Date _____

1. Determine the area and perimeter of the rectangle.

8 cm

2 cm

2. Determine the perimeter of the rectangle.

347 m

99 m

Lesson 1:   Investigate and use the formulas for area and perimeter of rectangles.

5

© 2018 Great Minds®. eureka-math.org

Tommy's dad is teaching him how to make tables out of tiles. Tommy makes a small table that is 3 feet wide and 4 feet long. How many square-foot tiles does he need to cover the top of the table? How many feet of decorative border material will his dad need to cover the edges of the table?

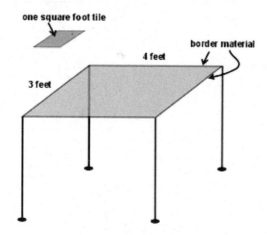

Extension: Tommy's dad is making a table 6 feet wide and 8 feet long. When both tables are placed together, what will their combined area be?

_____

_____

_____

_____

_____

Read   Draw   Write

Lesson 2:   Solve multiplicative comparison word problems by applying the area and perimeter formulas.

© 2018 Great Minds®. eureka-math.org

7

Name _____   Date _____

1. A rectangular porch is 4 feet wide. It is 3 times as long as it is wide.

    a.   Label the diagram with the dimensions of the porch.

    b.   Find the perimeter of the porch.

2. A narrow rectangular banner is 5 inches wide. It is 6 times as long as it is wide.

    a.   Draw a diagram of the banner, and label its dimensions.

    b.   Find the perimeter and area of the banner.

EUREKA
MATH®

Lesson 2:   Solve multiplicative comparison word problems by applying the area
            and perimeter formulas.

© 2018 Great Minds®. eureka-math.org

9

3. The area of a rectangle is 42 square centimeters. Its length is 7 centimeters.

   a. What is the width of the rectangle?

   b. Charlie wants to draw a second rectangle that is the same length but is 3 times as wide. Draw and label Charlie's second rectangle.

   c. What is the perimeter of Charlie's second rectangle?

Lesson 2:   Solve multiplicative comparison word problems by applying the area
              and perimeter formulas.

© 2018 Great Minds®. eureka-math.org

EUREKA
MATH

4. The area of Betsy's rectangular sandbox is 20 square feet. The longer side measures 5 feet. The sandbox at the park is twice as long and twice as wide as Betsy's.

   a. Draw and label a diagram of Betsy's sandbox. What is its perimeter?

   b. Draw and label a diagram of the sandbox at the park. What is its perimeter?

   c. What is the relationship between the two perimeters?

   d. Find the area of the park's sandbox using the formula A = l × w.

EUREKA MATH®

Lesson 2:   Solve multiplicative comparison word problems by applying the area
              and perimeter formulas.

© 2018 Great Minds®. eureka-math.org

11

e. The sandbox at the park has an area that is how many times that of Betsy's sandbox?

f. Compare how the perimeter changed with how the area changed between the two sandboxes. Explain what you notice using words, pictures, or numbers.

Lesson 2:   Solve multiplicative comparison word problems by applying the area and perimeter formulas.

EUREKA MATH

Name _____   Date _____

1. A table is 2 feet wide. It is 6 times as long as it is wide.

    a. Label the diagram with the dimensions of the table.

    b. Find the perimeter of the table.

2. A blanket is 4 feet wide. It is 3 times as long as it is wide.

    a. Draw a diagram of the blanket, and label its dimensions.

    b. Find the perimeter and area of the blanket.

Lesson 2:   Solve multiplicative comparison word problems by applying the area
            and perimeter formulas.

© 2018 Great Minds®. eureka-math.org

13

Name _____   Date _____

Solve the following problems. Use pictures, numbers, or words to show your work.

1. The rectangular projection screen in the school auditorium is 5 times as long and 5 times as wide as the
    rectangular screen in the library. The screen in the library is 4 feet long with a perimeter of 14 feet. What
    is the perimeter of the screen in the auditorium?

2. The width of David's rectangular tent is 5 feet. The length is twice the width. David's rectangular air
    mattress measures 3 feet by 6 feet. If David puts the air mattress in the tent, how many square feet of
    floor space will be available for the rest of his things?

3. Jackson's rectangular bedroom has an area of 90 square feet. The area of his bedroom is 9 times that of his rectangular closet. If the closet is 2 feet wide, what is its length?

4. The length of a rectangular deck is 4 times its width. If the deck's perimeter is 30 feet, what is the deck's area?

Lesson 3:   Demonstrate understanding of area and perimeter formulas by solving multi-step real-world problems.

EUREKA
MATH

Name _____   Date _____

Solve the following problem. Use pictures, numbers, or words to show your work.

A rectangular poster is 3 times as long as it is wide. A rectangular banner is 5 times as long as it is wide. Both the banner and the poster have perimeters of 24 inches. What are the lengths and widths of the poster and the banner?

Lesson 3:   Demonstrate understanding of area and perimeter formulas by solving multi-step real-world problems.

© 2018 Great Minds®. eureka-math.org

17

Samantha received an allowance of $3 every week. By babysitting, she earned an additional $30 every week. How much money did Samantha have in four weeks, combining her allowance and her babysitting?

Read   **Draw**   **Write**

EUREKA
MATH.

Lesson 4:   Interpret and represent patterns when multiplying by 10, 100, and
                   1,000 in arrays and numerically.

19

© 2018 Great Minds®. eureka-math.org

Name _____   Date _____

Example:

$5 \times 10 =$ ___**50**___

5 ones $\times 10 =$ ___**5 tens**___

thousands	hundreds	tens	ones
			●●●●●
		○○○○○	

Draw place value disks and arrows as shown to represent each product.

1. $5 \times 100 =$ _____

    $5 \times 10 \times 10 =$ _____

    5 ones $\times 100 =$ _____ _____

thousands	hundreds	tens	ones

2. $5 \times 1,000 =$ _____

    $5 \times 10 \times 10 \times 10 =$ _____

    5 ones $\times 1,000 =$ _____ _____

thousands	hundreds	tens	ones

3. Fill in the blanks in the following equations.

    a. $6 \times 10 =$ _____

    b. _____ $\times 6 = 600$

    c. $6,000 =$ _____ $\times 1,000$

    d. $10 \times 4 =$ _____

    e. $4 \times$ _____ $= 400$

    f. _____ $\times 4 = 4,000$

    g. $1,000 \times 9 =$ _____

    h. _____ $= 10 \times 9$

    i. $900 =$ _____ $\times 100$

Draw place value disks and arrows to represent each product.

4. 12 × 10 = _____

    (1 ten 2 ones) × 10 = _____

thousands	hundreds	tens	ones

5. 18 × 100 = _____

    18 × 10 × 10 = _____

    (1 ten 8 ones) × 100 = _____

thousands	hundreds	tens	ones

6. 25 × 1,000 = _____

    25 × 10 × 10 × 10 = _____

    (2 tens 5 ones) × 1,000 =

    _____

ten thousands	thousands	hundreds	tens	ones

Decompose each multiple of 10, 100, or 1,000 before multiplying.

7. 3 × 40 = 3 × 4 × _____

    = 12 × _____

    = _____

8. 3 × 200 = 3 × _____ × _____

    = _____ × _____

    = _____

9. 4 × 4,000 = _____ × _____ × _____

    = _____ × _____

    = _____

10. 5 × 4,000 = _____ × _____ × _____

    = _____ × _____

    = _____

Lesson 4:   Interpret and represent patterns when multiplying by 10, 100, and 1,000 in arrays and numerically.

© 2018 Great Minds®. eureka-math.org

EUREKA
MATH®

Name _____ Date _____

Fill in the blanks in the following equations.

a. 5 × 10 = _____

b. _____ × 5 = 500

c. 5,000 = _____ × 1000

d. 10 × 2 = _____

e. _____ × 20 = 2,000

f. 2,000 = 10 × _____

g. 100 × 18 = _____

h. _____ = 10 × 32

i. 4,800 = _____ × 100

j. 60 × 4 = _____

k. 5 × 600 = _____

l. 8,000 × 5 = _____

EUREKA
MATH®

Lesson 4: Interpret and represent patterns when multiplying by 10, 100, and 1,000 in arrays and numerically.

23

© 2018 Great Minds®. eureka-math.org

thousands	hundreds	tens	ones

thousands place value chart

Lesson 4:   Interpret and represent patterns when multiplying by 10, 100, and 1,000 in arrays and numerically.

© 2018 Great Minds®. eureka-math.org

25

Name _____   Date _____

Draw place value disks to represent the value of the following expressions.

1. $2 \times 3 =$ _____

    2 times _____ ones is _____ ones.

thousands	hundreds	tens	ones

    $$\begin{array}{r} 3 \\ \times\ 2 \\ \hline \end{array}$$

2. $2 \times 30 =$ _____

    2 times _____ tens is _____.

thousands	hundreds	tens	ones

    $$\begin{array}{r} 30 \\ \times\ \ 2 \\ \hline \end{array}$$

3. $2 \times 300 =$ _____

    2 times _____ is _____.

thousands	hundreds	tens	ones

    $$\begin{array}{r} 300 \\ \times\ \ \ 2 \\ \hline \end{array}$$

4. $2 \times 3,000 =$ _____

    _____ times _____ is _____.

thousands	hundreds	tens	ones

    $$\begin{array}{r} 3,000 \\ \times\ \ \ \ \ 2 \\ \hline \end{array}$$

Lesson 5:   Multiply multiples of 10, 100, and 1,000 by single digits, recognizing patterns.

© 2018 Great Minds®. eureka-math.org

27

5. Find the product.

a. 20×7	b. 3×60	c. 3×400	d. 2×800
e. 7×30	f. 60×6	g. 400×4	h. $4 \times 8,000$
i. 5×30	j. 5×60	k. 5×400	l. $8,000 \times 5$

6. Brianna buys 3 packs of balloons for a party. Each pack has 60 balloons. How many balloons does Brianna have?

Lesson 5:   Multiply multiples of 10, 100, and 1,000 by single digits, recognizing patterns.

© 2018 Great Minds®. eureka-math.org

7. Jordan has twenty times as many baseball cards as his brother. His brother has 9 cards. How many cards does Jordan have?

8. The aquarium has 30 times as many fish in one tank as Jacob has. The aquarium has 90 fish. How many fish does Jacob have?

Lesson 5:   Multiply multiples of 10, 100, and 1,000 by single digits, recognizing patterns.

© 2018 Great Minds®. eureka-math.org

29

Name _____   Date _____

Draw place value disks to represent the value of the following expressions.

1. $4 \times 200 =$ _____

4 times _____ is _____.

thousands	hundreds	tens	ones

$$\begin{array}{r} 200 \\ \times \quad 4 \\ \hline \end{array}$$

2. $4 \times 2{,}000 =$ _____

_____ times _____ is _____.

thousands	hundreds	tens	ones

$$\begin{array}{r} 2{,}000 \\ \times \quad 4 \\ \hline \end{array}$$

3. Find the product.

a. 30×3	b. 8×20	c. 6×400	d. 2×900
e. 8×80	f. 30×4	g. 500×6	h. $8 \times 5{,}000$

4. Bonnie worked for 7 hours each day for 30 days. How many hours did she work altogether?

EUREKA MATH®

Lesson 5: Multiply multiples of 10, 100, and 1,000 by single digits, recognizing patterns.

© 2018 Great Minds®. eureka-math.org

There are 400 children at Park Elementary School. Park High School has 4 times as many students.

a.   How many students in all attend both schools?

b.   Lane High School has 5 times as many students as Park Elementary. How many more students attend Lane High School than Park High School?

Read   **Draw**   **Write**

Lesson 6:   Multiply two-digit multiples of 10 by two-digit multiples of 10 with the area model.

© 2018 Great Minds®. eureka-math.org

33

Name _____   Date _____

Represent the following problem by drawing disks in the place value chart.

1. To solve 20 × 40, think

hundreds	tens	ones

    (2 tens × 4) × 10 = _____

    20 × (4 × 10) = _____

    20 × 40 = _____

2. Draw an area model to represent 20 × 40.

        2 tens × 4 tens = _____ _____

3. Draw an area model to represent 30 × 40.

        3 tens × 4 tens = _____ _____

    30 × 40 = _____

4. Draw an area model to represent 20 × 50.

    2 tens × 5 tens = _____ _____

    20 × 50 = _____

Rewrite each equation in unit form and solve.

5. 20 × 20 = _____

    2 tens × 2 tens = _____ hundreds

6. 60 × 20 = _____

    6 tens × 2 _____ = _____ hundreds

7. 70 × 20 = _____

    _____ tens × _____ tens = 14 _____

8. 70 × 30 = _____

    ____ _____ × ____ _____ = _____ hundreds

Lesson 6:   Multiply two-digit multiples of 10 by two-digit multiples of 10 with the area model.

EUREKA MATH®

9. If there are 40 seats per row, how many seats are in 90 rows?

10. One ticket to the symphony costs $50. How much money is collected if 80 tickets are sold?

Lesson 6:   Multiply two-digit multiples of 10 by two-digit multiples of 10 with the
              area model.

© 2018 Great Minds®. eureka-math.org

37

Name _____ Date _____

Represent the following problem by drawing disks in the place value chart.

1. To solve 20 × 30, think

   (2 tens × 3) × 10 = _____

   20 × (3 × 10) = _____

   20 × 30 = _____

hundreds	tens	ones

2. Draw an area model to represent 20 × 30.

   2 tens × 3 tens = _____ _____

3. Every night, Eloise reads 40 pages. How many total pages does she read at night during the 30 days of November?

EUREKA MATH

Lesson 6: Multiply two-digit multiples of 10 by two-digit multiples of 10 with the area model.

© 2018 Great Minds®. eureka-math.org

39

The basketball team is selling T-shirts for $9 each. On Monday, they sold 4 T-shirts. On Tuesday, they sold 5 times as many T-shirts as on Monday. How much money did the team earn altogether on Monday and Tuesday?

_____

_____

_____

_____

Read   **Draw**   **Write**

Name _____   Date _____

1. Represent the following expressions with disks, regrouping as necessary, writing a matching expression, and recording the partial products vertically as shown below.

    a. 1×43

tens	ones
● ● ● ●	● ● ●

$$
\begin{array}{r}
4\ 3 \\
\times\quad\ 1 \\
\hline
3 \\
+\ 4\ 0 \\
\hline
4\ 3
\end{array}
$$

→ 1×3 ones
→ 1×4 tens

    b. 2×43

tens	ones

    c. 3×43

hundreds	tens	ones

EUREKA
MATH

d.   4 × 43

hundreds	tens	ones

2.   Represent the following expressions with disks, regrouping as necessary. To the right, record the partial products vertically.

    a.   2 × 36

hundreds	tens	ones

    b.   3 × 61

hundreds	tens	ones

    c.   4 × 84

hundreds	tens	ones

Lesson 7:   Use place value disks to represent two-digit by one-digit multiplication.

EUREKA MATH

Name _____   Date _____

Represent the following expressions with disks, regrouping as necessary. To the right, record the partial products vertically.

1. 6 × 41

hundreds	tens	ones

2. 7 × 31

hundreds	tens	ones

ten thousands	thousands	hundreds	tens	ones

ten thousands place value chart

Lesson 7:   Use place value disks to represent two-digit by one-digit multiplication.

47

© 2018 Great Minds®. eureka-math.org

Andre buys a stamp to mail a letter. The stamp costs 46 cents. Andre also mails a package. The postage to mail the package costs 5 times as much as the cost of the stamp. How much does it cost to mail the package and letter?

_____

_____

_____

_____

Read   **Draw**   **Write**

EUREKA MATH®

Lesson 8:   Extend the use of place value disks to represent three- and four-digit by one-digit multiplication.

© 2018 Great Minds®. eureka-math.org

49

1. Represent the following expressions with disks, regrouping as necessary, writing a matching expression, and recording the partial products vertically as shown below.

    a. 1×213

hundreds	tens	ones

    $$\begin{array}{r} 2\quad 1\quad 3 \\ \times \qquad\quad 1 \\ \hline \end{array}$$

    → 1×3 ones
    → 1×1 ten
    → 1×2 hundreds

    $+$ _____

    $1 \times$ ___ hundreds $+ \; 1 \times$ ___ ten $+ \; 1 \times$ ___ ones

    b. 2×213

hundreds	tens	ones

    c. 3×214

hundreds	tens	ones

EUREKA MATH

Lesson 8:   Extend the use of place value disks to represent three- and four-digit by one-digit multiplication.

© 2018 Great Minds®. eureka-math.org

51

d. 3 × 1,254

thousands	hundreds	tens	ones

2. Represent the following expressions with disks, using either method shown during class, regrouping as necessary. To the right, record the partial products vertically.

a. 3 × 212

b. 2 × 4,036

Lesson 8:   Extend the use of place value disks to represent three- and four-digit by one-digit multiplication.

© 2018 Great Minds®. eureka-math.org

EUREKA MATH

c.   3 × 2,546

d.   3 × 1,407

3.   Every day at the bagel factory, Cyndi makes 5 different kinds of bagels. If she makes 144 of each kind, what is the total number of bagels that she makes?

Lesson 8:   Extend the use of place value disks to represent three- and four-digit by one-digit multiplication.

© 2018 Great Minds®. eureka-math.org

53

Name _____ Date _____

Represent the following expressions with disks, regrouping as necessary. To the right, record the partial products vertically.

1. 4 × 513

2. 3 × 1,054

Lesson 8: Extend the use of place value disks to represent three- and four-digit by one-digit multiplication.

© 2018 Great Minds®. eureka-math.org

55

Calculate the total amount of milk in three cartons if each carton contains 236 mL of milk.

_____

_____

_____

_____

Read   **Draw**   **Write**

Name _____   Date _____

1. Solve using each method.

Partial Products	Standard Algorithm

a.   3   4
    ×   4
	3 4
	    × 4

Partial Products	Standard Algorithm

b.   2   2   4
    ×   3
	2 2 4
	    × 3

2. Solve. Use the standard algorithm.

a.
 	
     2   5   1
 	
    ×   3
	b.
	
	     1 3 5
	
	    × 6
		c.
		
		     3 0 4
		
		    × 9

d.
 	
     4   0   5
 	
    ×   4
	e.
	
	     3 1 6
	
	    × 5
		f.
		
		     3 9 2
		
		    × 6

EUREKA MATH®

Lesson 9:   Multiply three- and four-digit numbers by one-digit numbers applying the standard algorithm.

© 2018 Great Minds®. eureka-math.org

59

3. The product of 7 and 86 is _____.

4. 9 times as many as 457 is _____.

5. Jashawn wants to make 5 airplane propellers.
    He needs 18 centimeters of wood for each propeller.
    How many centimeters of wood will he use?

Lesson 9:   Multiply three- and four-digit numbers by one-digit numbers applying the standard algorithm.

© 2018 Great Minds®. eureka-math.org

6. One game system costs $238. How much will 4 game systems cost?

7. A small bag of chips weighs 48 grams. A large bag of chips weighs three times as much as the small bag. How much will 7 large bags of chips weigh?

EUREKA
MATH®

Lesson 9:   Multiply three- and four-digit numbers by one-digit numbers applying the standard algorithm.

© 2018 Great Minds®. eureka-math.org

61

Name _____   Date _____

1. Solve using the standard algorithm.

a.	b.
$\begin{array}{r} 6\ \ 0\ \ 8 \\ \times \qquad 9 \\ \hline \end{array}$	$\begin{array}{r} 5\ \ 7\ \ 4 \\ \times \qquad 7 \\ \hline \end{array}$

2. Morgan is 23 years old. Her grandfather is 4 times as old. How old is her grandfather?

EUREKA MATH

Lesson 9: Multiply three- and four-digit numbers by one-digit numbers applying the standard algorithm.

© 2018 Great Minds®. eureka-math.org

63

The principal wants to buy 8 pencils for every student at her school. If there are 859 students, how many pencils does the principal need to buy?

Read   **Draw**   **Write**

EUREKA MATH

Lesson 10:   Multiply three- and four-digit numbers by one-digit numbers applying the standard algorithm.

© 2018 Great Minds®. eureka-math.org

65

Name _____ Date _____

1. Solve using the standard algorithm.

a. 3 × 42	b. 6 × 42
c. 6 × 431	d. 3 × 431
e. 3 × 6,212	f. 3 × 3,106
g. 4 × 4,309	h. 4 × 8,618

EUREKA MATH®

Lesson 10:   Multiply three- and four-digit numbers by one-digit numbers applying the standard algorithm.

67

© 2018 Great Minds®. eureka-math.org

2. There are 365 days in a common year. How many days are in 3 common years?

3. The length of one side of a square city block is 462 meters. What is the perimeter of the block?

4. Jake ran 2 miles. Jesse ran 4 times as far. There are 5,280 feet in a mile. How many feet did Jesse run?

Lesson 10:   Multiply three- and four-digit numbers by one-digit numbers
applying the standard algorithm.

EUREKA
MATH®

Name _____ Date _____

1. Solve using the standard algorithm.

a. 2,348 × 6	b. 1,679 × 7

2. A farmer planted 4 rows of sunflowers. There were 1,205 plants in each row. How many sunflowers did he plant?

EUREKA MATH

Lesson 10: Multiply three- and four-digit numbers by one-digit numbers applying the standard algorithm.

© 2018 Great Minds®. eureka-math.org

69

Write an equation for the area of each rectangle. Then, find the sum of the two areas.

Extension: Find a faster method for finding the area of the combined rectangles.

_____

_____

_____

_____

Read   Draw   Write

Name _____ Date _____

1. Solve the following expressions using the standard algorithm, the partial products method, and the area model.

a. 4 2 5 × 4

4 (400 + 20 + 5)

(4 × _____) + (4 × _____) + (4 × _____)

b. 5 3 4 × 7

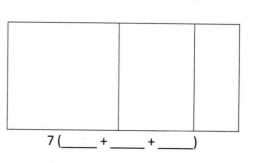

7 (_____ + _____ + _____)

(__ × _____) + (__ × _____) + (__ × _____)

c. 2 0 9 × 8

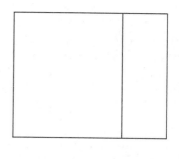

__ (_____ + _____)

(__ × _____) + (__ × _____)

EUREKA MATH

Lesson 11: Connect the area model and the partial products method to the standard algorithm.

73

© 2018 Great Minds®. eureka-math.org

2. Solve using the partial products method.

Cayla's school has 258 students. Janet's school has 3 times as many students as Cayla's. How many students are in Janet's school?

3. Model with a tape diagram and solve.

4 times as much as 467

Solve using the standard algorithm, the area model, the distributive property, or the partial products method.

4. $5,131 \times 7$

Lesson 11:   Connect the area model and the partial products method to the standard algorithm.

EUREKA
MATH®

5. 3 times as many as 2,805

6. A restaurant sells 1,725 pounds of spaghetti and 925 pounds of linguini every month. After 9 months, how many pounds of pasta does the restaurant sell?

Name _____   Date _____

1. Solve using the standard algorithm, the area model, the distributive property, or the partial products method.

    2,809 × 4

2. The monthly school newspaper is 9 pages long. Mrs. Smith needs to print 675 copies. What will be the total number of pages printed?

EUREKA MATH

Lesson 11:   Connect the area model and the partial products method to the standard algorithm.

© 2018 Great Minds®. eureka-math.org

77

Name _____ Date _____

Use the RDW process to solve the following problems.

1. The table shows the cost of party favors. Each party
    guest receives a bag with 1 balloon, 1 lollipop, and 1 bracelet.
    What is the total cost for 9 guests?

Item	Cost
1 balloon	26¢
1 lollipop	14¢
1 bracelet	33¢

2. The Turner family uses 548 liters of water per day. The Hill family uses 3 times as much water per day.
    How much water does the Hill family use per week?

3. Jayden has 347 marbles. Elvis has 4 times as many as Jayden. Presley has 799 fewer than Elvis.
    How many marbles does Presley have?

Lesson 12:   Solve two-step word problems, including multiplicative comparison.

79

© 2018 Great Minds®. eureka-math.org

4. a. Write an equation that would allow someone to find the value of R.

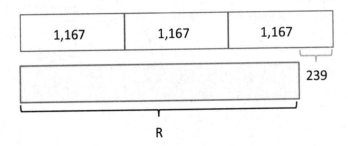

b. Write your own word problem to correspond to the tape diagram, and then solve.

Lesson 12:   Solve two-step word problems, including multiplicative comparison.

EUREKA
MATH

Name _____   Date _____

Use the RDW process to solve the following problem.

Jennifer has 256 beads. Stella has 3 times as many beads as Jennifer. Tiah has 104 more beads than Stella. How many beads does Tiah have?

Lesson 12:   Solve two-step word problems, including multiplicative comparison.

81

© 2018 Great Minds®. eureka-math.org

Name _____   Date _____

Solve using the RDW process.

1. Over the summer, Kate earned $180 each week for 7 weeks. Of that money, she spent $375 on a new computer and $137 on new clothes. How much money did she have left?

2. Sylvia weighed 8 pounds when she was born. By her first birthday, her weight had tripled. By her second birthday, she had gained 12 more pounds. At that time, Sylvia's father weighed 5 times as much as she did. What was Sylvia and her dad's combined weight?

Lesson 13:   Use multiplication, addition, or subtraction to solve multi-step word problems.

3. Three boxes weighing 128 pounds each and one box weighing 254 pounds were loaded onto the back of an empty truck. A crate of apples was then loaded onto the same truck. If the total weight loaded onto the truck was 2,000 pounds, how much did the crate of apples weigh?

4. In one month, Charlie read 814 pages. In the same month, his mom read 4 times as many pages as Charlie, and that was 143 pages more than Charlie's dad read. What was the total number of pages read by Charlie and his parents?

Lesson 13:   Use multiplication, addition, or subtraction to solve multi-step word problems.

EUREKA
MATH®

Name _____   Date _____

Solve using the RDW process.

1. Michael earns $9 per hour. He works 28 hours each week. How much does he earn in 6 weeks?

2. David earns $8 per hour. He works 40 hours each week. How much does he earn in 6 weeks?

3. After 6 weeks, who earned more money? How much more money?

Lesson 13:   Use multiplication, addition, or subtraction to solve multi-step word
               problems.

© 2018 Great Minds®. eureka-math.org

85

Tyler planted potatoes, oats, and corn. He planted 23 acres of potatoes. He planted 3 times as many acres of oats as potatoes, and he planted 4 times as many acres of corn as oats. How many acres did Tyler plant with potatoes, oats, and corn in all?

_____

_____

_____

_____

Read   **Draw**   **Write**

Lesson 14:   Solve division word problems with remainders.

87

© 2018 Great Minds®. eureka-math.org

Name _____   Date _____

Use the RDW process to solve the following problems.

1. There are 19 identical socks. How many pairs of socks are there? Will there be any socks without a match? If so, how many?

2. If it takes 8 inches of ribbon to make a bow, how many bows can be made from 3 feet of ribbon (1 foot = 12 inches)? Will any ribbon be left over? If so, how much?

3. The library has 27 chairs and 5 tables. If the same number of chairs is placed at each table, how many chairs can be placed at each table? Will there be any extra chairs? If so, how many?

4. The baker has 42 kilograms of flour. She uses 8 kilograms each day. After how many days will she need to buy more flour?

5. Caleb has 76 apples. He wants to bake as many pies as he can. If it takes 8 apples to make each pie, how many apples will he use? How many apples will not be used?

6. Forty-five people are going to the beach. Seven people can ride in each van. How many vans will be required to get everyone to the beach?

EUREKA
MATH

Name _____   Date _____

Use the RDW process to solve the following problem.

Fifty-three students are going on a field trip. The students are divided into groups of 6 students. How many groups of 6 students will there be? If the remaining students form a smaller group, and one chaperone is assigned to every group, how many total chaperones are needed?

Chandra printed 38 photos to put into her scrapbook. If she can fit 4 photos on each page, how many pages will she use for her photos?

_____

_____

_____

_____

Read   **Draw**   **Write**

EUREKA MATH®

Lesson 15:   Understand and solve division problems with a remainder using the array and area models.

93

© 2018 Great Minds®. eureka-math.org

Name _____   Date _____

Show division using an array.	Show division using an area model.
1. $18 \div 6$ Quotient = _____ Remainder = _____	 Can you show $18 \div 6$ with one rectangle? _____
2. $19 \div 6$ Quotient = _____ Remainder = _____	 Can you show $19 \div 6$ with one rectangle? _____ Explain how you showed the remainder:

EUREKA MATH

Lesson 15:   Understand and solve division problems with a remainder using the array and area models.

© 2018 Great Minds®. eureka-math.org

95

Solve using an array and an area model. The first one is done for you.

Example: 25 ÷ 2

a.

Quotient = 12   Remainder = 1

b.

3. 29 ÷ 3

a.

b.

4. 22 ÷ 5

a.

b.

5. 43 ÷ 4

a.

b.

6. 59 ÷ 7

a.

b.

Lesson 15:   Understand and solve division problems with a remainder using the array and area models.

EUREKA
MATH

Name _____   Date _____

Solve using an array and area model.

1.   27 ÷ 5

    a.   b.

2.   32 ÷ 6

    a.   b.

Lesson 15:   Understand and solve division problems with a remainder using the
               array and area models.

97

© 2018 Great Minds®. eureka-math.org

Name _____ Date _____

Show the division using disks. Relate your work on the place value chart to long division. Check your quotient and remainder by using multiplication and addition.

1. $7 \div 2$

Ones

$2 \overline{\smash{)}7}$

Check Your Work

$$\begin{array}{r} 3 \\ \times\ 2 \\ \hline \end{array}$$

quotient = _____

remainder = _____

2. $27 \div 2$

Tens	Ones

$2 \overline{\smash{)}27}$

Check Your Work

quotient = _____

remainder = _____

EUREKA MATH

Lesson 16: Understand and solve two-digit dividend division problems with a remainder in the ones place by using place value disks.

99

3.   8 ÷ 3

Ones

$3 \overline{)\ 8\ }$

quotient = _____

remainder = _____

Check Your Work

4.   38 ÷ 3

Tens	Ones

$3 \overline{)\ 38\ }$

Check Your Work

quotient = _____

remainder = _____

Lesson 16:   Understand and solve two-digit dividend division problems with a
remainder in the ones place by using place value disks.

EUREKA
MATH®

5. 6 ÷ 4

Ones

4 ⟌ 6

quotient = _____

remainder = _____

Check Your Work

6. 86 ÷ 4

Tens	Ones

4 ⟌ 86

Check Your Work

quotient = _____

remainder = _____

Lesson 16:   Understand and solve two-digit dividend division problems with a
remainder in the ones place by using place value disks.

101

© 2018 Great Minds®. eureka-math.org

Name _____ Date _____

Show the division using disks. Relate your work on the place value chart to long division. Check your quotient and remainder by using multiplication and addition.

1. 5 ÷ 3

Ones

3 ⟌ 5

Check Your Work

quotient = _____

remainder = _____

2. 65 ÷ 3

Tens	Ones

3 ⟌ 6 5

Check Your Work

quotient = _____

remainder = _____

Lesson 16: Understand and solve two-digit dividend division problems with a remainder in the ones place by using place value disks.

103

EUREKA MATH

tens	ones

tens place value chart

Lesson 16:   Understand and solve two-digit dividend division problems with a remainder in the ones place by using place value disks.

© 2018 Great Minds®. eureka-math.org

105

Audrey and her sister found 9 dimes and 8 pennies. If they share the money equally, how much money will each sister get?

_____

_____

_____

_____

Read   **Draw**   **Write**

Lesson 17:   Represent and solve division problems requiring decomposing a
               remainder in the tens.

© 2018 Great Minds®. eureka-math.org

107

Name _____   Date _____

Show the division using disks. Relate your model to long division. Check your quotient and remainder by using multiplication and addition.

1. 5 ÷ 2

Ones

2 ⟌ 5

Check Your Work

$$\begin{array}{r} 2 \\ \times\ 2 \\ \hline \end{array}$$

quotient = _____

remainder = _____

2. 50 ÷ 2

Tens	Ones

2 ⟌ 50

Check Your Work

quotient = _____

remainder = _____

EUREKA MATH

3. 7 ÷ 3

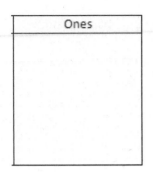

Ones

3 ⟌ 7

Check Your Work

quotient = _____

remainder = _____

4. 75 ÷ 3

Tens	Ones

3 ⟌ 7 5

Check Your Work

quotient = _____

remainder = _____

Lesson 17:   Represent and solve division problems requiring decomposing a remainder in the tens.

EUREKA MATH

5. 9 ÷ 4

Ones

4 ⟌ 9

Check Your Work

quotient = _____

remainder = _____

6. 92 ÷ 4

Tens	Ones

4 ⟌ 92

Check Your Work

quotient = _____

remainder = _____

EUREKA MATH

Lesson 17:   Represent and solve division problems requiring decomposing a remainder in the tens.

© 2018 Great Minds®. eureka-math.org

111

Name _____ Date _____

Show the division using disks. Relate your model to long division. Check your quotient by using multiplication and addition.

1. $5 \div 4$

Ones

$4\overline{)5}$

Check Your Work

quotient = _____

remainder = _____

2. $56 \div 4$

Tens	Ones

$4\overline{)56}$

Check Your Work

quotient = _____

remainder = _____

EUREKA MATH

Lesson 17:   Represent and solve division problems requiring decomposing a remainder in the tens.

113

© 2018 Great Minds®. eureka-math.org

Malory's family is going to buy oranges. The Grand Market sells oranges at 3 pounds for 87 cents.

How much does 1 pound of oranges cost at Grand Market?

_____

_____

_____

_____

Read   **Draw**   **Write**

Name _____ Date _____

Solve using the standard algorithm. Check your quotient and remainder by using multiplication and addition.

1. 46 ÷ 2	2. 96 ÷ 3
3. 85 ÷ 5	4. 52 ÷ 4
5. 53 ÷ 3	6. 95 ÷ 4

7. $89 \div 6$	8. $96 \div 6$
9. $60 \div 3$	10. $60 \div 4$
11. $95 \div 8$	12. $95 \div 7$

Lesson 18:   Find whole number quotients and remainders.

EUREKA MATH®

Name _____ Date _____

Solve using the standard algorithm. Check your quotient and remainder by using multiplication and addition.

1. 93 ÷ 7

2. 99 ÷ 8

Two friends start a business writing and selling comic books. After 1 month, they have earned $38.

Show how they can share their earnings fairly, using $1, $5, $10, and $20 bills.

Read   **Draw**   **Write**

Name _____ Date _____

1. When you divide 94 by 3, there is a remainder of 1. Model this problem with place value disks. In the place value disk model, how did you show the remainder?

2. Cayman says that 94 ÷ 3 is 30 with a remainder of 4. He reasons this is correct because (3 × 30) + 4 = 94. What mistake has Cayman made? Explain how he can correct his work.

Lesson 19:   Explain remainders by using place value understanding and models.

123

© 2018 Great Minds®. eureka-math.org

3. The place value disk model is showing 72 ÷ 3.
    Complete the model. Explain what happens to the
    1 ten that is remaining in the tens column.

4. Two friends evenly share 56 dollars.

    a. They have 5 ten-dollar bills and 6 one-dollar bills. Draw a picture to show how the bills will be shared.
        Will they have to make change at any stage?

    b. Explain how they share the money evenly.

       Lesson 19:   Explain remainders by using place value understanding and models.

EUREKA
MATH

5. Imagine you are filming a video explaining the problem 45 ÷ 3 to new fourth graders. Create a script to explain how you can keep dividing after getting a remainder of 1 ten in the first step.

Lesson 19: Explain remainders by using place value understanding and models.

125

© 2018 Great Minds®. eureka-math.org

Name _____   Date _____

1. Molly's photo album has a total of 97 pictures. Each page of the album holds 6 pictures. How many pages can Molly fill? Will there be any pictures left? If so, how many? Use place value disks to solve.

2. Marti's photo album has a total of 45 pictures. Each page holds 4 pictures. She said she can only fill 10 pages completely. Do you agree? Explain why or why not.

EUREKA MATH

Lesson 19:   Explain remainders by using place value understanding and models.

127

© 2018 Great Minds®. eureka-math.org

Write an expression to find the unknown length of each rectangle. Then, find the sum of the two unknown lengths.

a. 4 cm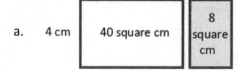

b. 4 cm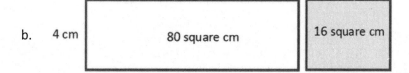

Read   Draw   Write

Lesson 20:   Solve division problems without remainders using the area model.

129

© 2018 Great Minds®. eureka-math.org

Name _____   Date _____

1. Alfonso solved a division problem by drawing an area model.

    a. Look at the area model. What division problem did Alfonso solve?

    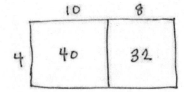

    b. Show a number bond to represent Alfonso's area model. Start with the total, and then show how the total is split into two parts. Below the two parts, represent the total length using the distributive property, and then solve.

    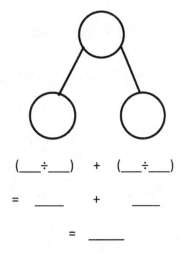

    (__÷__)   +   (__÷__)

    =   ____   +   ____

    =   _____

2. Solve 45 ÷ 3 using an area model. Draw a number bond, and use the distributive property to solve for the unknown length.

3. Solve 64 ÷ 4 using an area model. Draw a number bond to show how you partitioned the area, and represent the division with a written method.

4. Solve 92 ÷ 4 using an area model. Explain, using words, pictures, or numbers, the connection of the distributive property to the area model.

5. Solve 72 ÷ 6 using an area model and the standard algorithm.

EUREKA
MATH

Name _____   Date _____

1. Tony drew the following area model to find an unknown length. What division equation did he model?

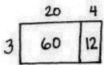

2. Solve 42 ÷ 3 using the area model, a number bond, and a written method.

Lesson 20:   Solve division problems without remainders using the area model.

© 2018 Great Minds®. eureka-math.org

133

A rectangle has an area of 36 square units and a width of 2 units. What is the unknown side length?

_____

_____

_____

Read   **Draw**   **Write**

Name _____ Date _____

1. Solve 37 ÷ 2 using an area model. Use long division and the distributive property to record your work.

2. Solve 76 ÷ 3 using an area model. Use long division and the distributive property to record your work.

3. Carolina solved the following division problem by drawing an area model.

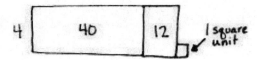

a. What division problem did she solve?

b. Show how Carolina's model can be represented using the distributive property.

Lesson 21: Solve division problems with remainders using the area model.

137

© 2018 Great Minds®. eureka-math.org

Solve the following problems using the area model. Support the area model with long division or the distributive property.

4. 48 ÷ 3	5. 49 ÷ 3
6. 56 ÷ 4	7. 58 ÷ 4
8. 66 ÷ 5	9. 79 ÷ 3

Lesson 21:   Solve division problems with remainders using the area model.

10. Seventy-three students are divided into groups of 6 students each. How many groups of 6 students are there? How many students will not be in a group of 6?

Lesson 21:   Solve division problems with remainders using the area model.

© 2018 Great Minds®. eureka-math.org

139

Name _____   Date _____

1. Kyle drew the following area model to find an unknown length. What division equation did he model?

$$2 \quad \boxed{\quad 40 \quad | \quad 18 \quad} \leftarrow \text{1 square unit}$$

2. Solve 93 ÷ 4 using the area model, long division, and the distributive property.

8 × _____ = 96. Find the unknown side length, or factor. Use an area model to solve the problem.

_____

_____

_____

_____

Read   **Draw**   **Write**

Lesson 22:   Find factor pairs for numbers to 100, and use understanding of factors
to define prime and composite.

143

© 2018 Great Minds®. eureka-math.org

Name _____ Date _____

1. Record the factors of the given numbers as multiplication sentences and as a list in order from least to greatest. Classify each as prime (P) or composite (C). The first problem is done for you.

	Multiplication Sentences	Factors	P or C
a.	4 1 × 4 = 4 2 × 2 = 4	The factors of 4 are: 1, 2, 4	C
b.	6	The factors of 6 are:	
c.	7	The factors of 7 are:	
d.	9	The factors of 9 are:	
e.	12	The factors of 12 are:	
f.	13	The factors of 13 are:	
g.	15	The factors of 15 are:	
h.	16	The factors of 16 are:	
i.	18	The factors of 18 are:	
j.	19	The factors of 19 are:	
k.	21	The factors of 21 are:	
l.	24	The factors of 24 are:	

EUREKA MATH

Lesson 22:   Find factor pairs for numbers to 100, and use understanding of factors to define prime and composite.

© 2018 Great Minds®. eureka-math.org

145

2. Find all factors for the following numbers, and classify each number as prime or composite. Explain your classification of each as prime or composite.

Factor Pairs for 25	

Factor Pairs for 28	

Factor Pairs for 29	

3. Bryan says all prime numbers are odd numbers.

   a. List all of the prime numbers less than 20 in numerical order.

   b. Use your list to show that Bryan's claim is false.

4. Sheila has 28 stickers to divide evenly among 3 friends. She thinks there will be no leftovers. Use what you know about factor pairs to explain if Sheila is correct.

EUREKA
MATH

Name _____   Date _____

Record the factors of the given numbers as multiplication sentences and as a list in order from least to greatest. Classify each as prime (P) or composite (C).

	Multiplication Sentences	Factors	Prime (P) or Composite (C)
a.	9	The factors of 9 are:	
b.	12	The factors of 12 are:	
c.	19	The factors of 19 are:	

Lesson 22: Find factor pairs for numbers to 100, and use understanding of factors to define prime and composite.

© 2018 Great Minds®. eureka-math.org

147

Sasha says that every number in the twenties is a composite number because 2 is even. Amanda says there are two prime numbers in the twenties. Who is correct? How do you know?

_____

_____

_____

_____

Read   **Draw**   **Write**

Lesson 23:   Use division and the associative property to test for factors and observe patterns.

149

EUREKA MATH

Name _____   Date _____

1. Explain your thinking or use division to answer the following.

a. Is 2 a factor of 84?	b. Is 2 a factor of 83?
c. Is 3 a factor of 84?	d. Is 2 a factor of 92?
e. Is 6 a factor of 84?	f. Is 4 a factor of 92?
g. Is 5 a factor of 84?	h. Is 8 a factor of 92?

EUREKA
MATH

Lesson 23:   Use division and the associative property to test for factors and observe patterns.

© 2018 Great Minds®. eureka-math.org

151

2. Use the associative property to find more factors of 24 and 36.

a. $24 = 12 \times 2$

$= (\underline{\hspace{1cm}} \times 3) \times 2$

$= \underline{\hspace{1cm}} \times (3 \times 2)$

$= \underline{\hspace{1cm}} \times 6$

$= \underline{\hspace{1cm}}$

b. $36 = \underline{\hspace{1cm}} \times 4$

$= (\underline{\hspace{1cm}} \times 3) \times 4$

$= \underline{\hspace{1cm}} \times (3 \times 4)$

$= \underline{\hspace{1cm}} \times 12$

$= \underline{\hspace{1cm}}$

3. In class, we used the associative property to show that when 6 is a factor, then 2 and 3 are factors, because $6 = 2 \times 3$. Use the fact that $8 = 4 \times 2$ to show that 2 and 4 are factors of 56, 72, and 80.

$$56 = 8 \times 7 \qquad\qquad 72 = 8 \times 9 \qquad\qquad 80 = 8 \times 10$$

4. The first statement is false. The second statement is true. Explain why, using words, pictures, or numbers.

   If a number has 2 and 4 as factors, then it has 8 as a factor.
   If a number has 8 as a factor, then both 2 and 4 are factors.

Lesson 23:   Use division and the associative property to test for factors and observe patterns.

© 2018 Great Minds®. eureka-math.org

EUREKA MATH

Name _____ Date _____

1. Explain your thinking or use division to answer the following.

a. Is 2 a factor of 34?	b. Is 3 a factor of 34?
c. Is 4 a factor of 72?	d. Is 3 a factor of 72?

2. Use the associative property to explain why the following statement is true.
   Any number that has 9 as a factor also has 3 as a factor.

Lesson 23:   Use division and the associative property to test for factors and observe patterns.

153

8 cm × 12 cm = 96 square centimeters. Imagine a rectangle with an area of 96 square centimeters and a side length of 4 centimeters. What is the length of its unknown side? How will it look when compared to the 8 centimeter by 12 centimeter rectangle? Draw and label both rectangles.

_____

_____

_____

_____

Read   **Draw**   **Write**

Name _____   Date _____

1. For each of the following, time yourself for 1 minute. See how many multiples you can write.

    a. Write the multiples of 5 starting from 100.

    b. Write the multiples of 4 starting from 20.

    c. Write the multiples of 6 starting from 36.

2. List the numbers that have 24 as a multiple.

3. Use mental math, division, or the associative property to solve. (Use scratch paper if you like.)

    a. Is 12 a multiple of 4? _____   Is 4 a factor of 12? _____

    b. Is 42 a multiple of 8? _____   Is 8 a factor of 42? _____

    c. Is 84 a multiple of 6? _____   Is 6 a factor of 84? _____

4. Can a prime number be a multiple of any other number except itself? Explain why or why not.

5. Follow the directions below.

1	2	3	4	5	6	7	8	9	10
11	12	13	14	15	16	17	18	19	20
21	22	23	24	25	26	27	28	29	30
31	32	33	34	35	36	37	38	39	40
41	42	43	44	45	46	47	48	49	50
51	52	53	54	55	56	57	58	59	60
61	62	63	64	65	66	67	68	69	70
71	72	73	74	75	76	77	78	79	80
81	82	83	84	85	86	87	88	89	90
91	92	93	94	95	96	97	98	99	100

a. Circle in red the multiples of 2. When a number is a multiple of 2, what are the possible values for the ones digit?

b. Shade in green the multiples of 3. Choose one. What do you notice about the sum of the digits? Choose another. What do you notice about the sum of the digits?

c. Circle in blue the multiples of 5. When a number is a multiple of 5, what are the possible values for the ones digit?

d. Draw an X over the multiples of 10. What digit do all multiples of 10 have in common?

Lesson 24:   Determine if a whole number is a multiple of another number.

© 2018 Great Minds®. eureka-math.org

EUREKA
MATH

Name _____   Date _____

1. Fill in the unknown multiples of 11.

    5 × 11 = _____

    6 × 11 = _____

    7 × 11 = _____

    8 × 11 = _____

    9 × 11 = _____

2. Complete the pattern of multiples by skip-counting.

    7, 14 _____, 28 _____, _____, _____, _____, _____, _____

3. a. List the numbers that have 18 as a multiple.

    b. What are the factors of 18?

    c. Are your two lists the same? Why or why not?

Name _____ Date _____

1. Follow the directions.

    Shade the number 1 red.

    a. Circle the first unmarked number.

    b. Cross off every multiple of that number except the one you circled. If it's already crossed off, skip it.

    c. Repeat Steps (a) and (b) until every number is either circled or crossed off.

    d. Shade every crossed out number in orange.

1	2	3	4	5	6	7	8	9	10
11	12	13	14	15	16	17	18	19	20
21	22	23	24	25	26	27	28	29	30
31	32	33	34	35	36	37	38	39	40
41	42	43	44	45	46	47	48	49	50
51	52	53	54	55	56	57	58	59	60
61	62	63	64	65	66	67	68	69	70
71	72	73	74	75	76	77	78	79	80
81	82	83	84	85	86	87	88	89	90
91	92	93	94	95	96	97	98	99	100

Lesson 25:   Explore properties of prime and composite numbers to 100 by using multiples.

© 2018 Great Minds®. eureka-math.org

161

2. a. List the circled numbers.

b. Why were the circled numbers not crossed off along the way?

c. Except for the number 1, what is similar about all of the numbers that were crossed off?

d. What is similar about all of the numbers that were circled?

Lesson 25: Explore properties of prime and composite numbers to 100 by using multiples.

© 2018 Great Minds®. eureka-math.org

EUREKA MATH

Name _____   Date _____

Use the calendar below to complete the following:

1. Cross off all composite numbers.

2. Circle all of the prime numbers.

3. List any remaining numbers.

Sunday	Monday	Tuesday	Wednesday	Thursday	Friday	Saturday
					1	2
3	4	5	6	7	8	9
10	11	12	13	14	15	16
17	18	19	20	21	22	23
24	25	26	27	28	29	30
31						

A coffee shop uses 8-ounce mugs to make all of its coffee drinks. In one week, they served 30 mugs of espresso, 400 lattes, and 5,000 mugs of coffee. How many ounces of coffee drinks did they make in that one week?

_____

_____

_____

_____

Read   **Draw**   **Write**

Name _____   Date _____

1. Draw place value disks to represent the following problems. Rewrite each in unit form and solve.

   a. 6 ÷ 2 = _____   ① ① ①   ① ① ①

       6 ones ÷ 2 = _____ ones

   b. 60 ÷ 2 = _____

       6 tens ÷ 2 = _____

   c. 600 ÷ 2 = _____

       _____ ÷ 2 = _____

   d. 6,000 ÷ 2 = _____

       _____ ÷ 2 = _____

2. Draw place value disks to represent each problem. Rewrite each in unit form and solve.

   a. 12 ÷ 3 = _____

       12 ones ÷ 3 = _____ ones

   b. 120 ÷ 3 = _____

       _____ ÷ 3 = _____

   c. 1,200 ÷ 3 = _____

       _____ ÷ 3 = _____

3. Solve for the quotient. Rewrite each in unit form.

a. 800 ÷ 2 = 400 8 hundreds ÷ 2 = 4 hundreds	b. 600 ÷ 2 = _____	c. 800 ÷ 4 = _____	d. 900 ÷ 3 = _____
e. 300 ÷ 6 = _____ 30 tens ÷ 6 = _____ tens	f. 240 ÷ 4 = _____	g. 450 ÷ 5 = _____	h. 200 ÷ 5 = _____
i. 3,600 ÷ 4 = _____ 36 hundreds ÷ 4 = _____ hundreds	j. 2,400 ÷ 4 = _____	k. 2,400 ÷ 3 = _____	l. 4,000 ÷ 5 = _____

4. Some sand weighs 2,800 kilograms. It is divided equally among 4 trucks. How many kilograms of sand are in each truck?

EUREKA MATH

5. Ivy has 5 times as many stickers as Adrian has. Ivy has 350 stickers. How many stickers does Adrian have?

6. An ice cream stand sold $1,600 worth of ice cream on Saturday, which was 4 times the amount sold on Friday. How much money did the ice cream stand collect on Friday?

EUREKA
MATH

Lesson 26:   Divide multiples of 10, 100, and 1,000 by single-digit numbers.

169

© 2018 Great Minds®. eureka-math.org

Name _____   Date _____

1. Solve for the quotient. Rewrite each in unit form.

a. $600 \div 3 = 200$ 6 hundreds $\div$ 3 = _____ hundreds	b. $1,200 \div 6 =$ _____	c. $2,100 \div 7 =$ _____	d. $3,200 \div 8 =$ _____

2. Hudson and 7 of his friends found a bag of pennies. There were 320 pennies, which they shared equally. How many pennies did each person get?

ones	tens	hundreds	thousands

thousands place value chart for dividing

Lesson 26: Divide multiples of 10, 100, and 1,000 by single-digit numbers.

173

© 2018 Great Minds®. eureka-math.org

Emma takes 57 stickers from her collection and divides them up equally between 4 of her friends. How many stickers will each friend receive? Emma puts the remaining stickers back in her collection. How many stickers will Emma return to her collection?

_____

_____

_____

Read   **Draw**   **Write**

Lesson 27:   Represent and solve division problems with up to a three-digit dividend numerically and with place value disks requiring decomposing a remainder in the hundreds place.

175

© 2018 Great Minds®. eureka-math.org

Name _____   Date _____

1. Divide. Use place value disks to model each problem.

---

a. 324 ÷ 2

---

b. 344 ÷ 2

Lesson 27: Represent and solve division problems with up to a three-digit
dividend numerically and with place value disks requiring
decomposing a remainder in the hundreds place.

177

c. $483 \div 3$

d. $549 \div 3$

Lesson 27:   Represent and solve division problems with up to a three-digit
dividend numerically and with place value disks requiring
decomposing a remainder in the hundreds place.

EUREKA
MATH®

2. Model using place value disks and record using the algorithm.

a. 655 ÷ 5
    Disks   Algorithm

b. 726 ÷ 3
    Disks   Algorithm

c. 688 ÷ 4
    Disks   Algorithm

Lesson 27:   Represent and solve division problems with up to a three-digit
              dividend numerically and with place value disks requiring
              decomposing a remainder in the hundreds place.

© 2018 Great Minds®. eureka-math.org

179

Name _____   Date _____

Divide. Use place value disks to model each problem. Then, solve using the algorithm.

1. 423 ÷ 3

Disks   Algorithm

2. 564 ÷ 4

Disks   Algorithm

Lesson 27:   Represent and solve division problems with up to a three-digit
                 dividend numerically and with place value disks requiring
                 decomposing a remainder in the hundreds place.

© 2018 Great Minds®. eureka-math.org

181

Use 846 ÷ 2 to write a word problem. Then, draw an accompanying tape diagram and solve.

_____

_____

_____

Read   **Draw**   **Write**

Lesson 28:   Represent and solve three-digit dividend division with divisors
of 2, 3, 4, and 5 numerically.

183

Name _____   Date _____

1. Divide. Check your work by multiplying. Draw disks on a place value chart as needed.

---

a. 574 ÷ 2

---

b. 861 ÷ 3

---

c. 354 ÷ 2

Lesson 28:   Represent and solve three-digit dividend division with divisors
              of 2, 3, 4, and 5 numerically.

185

© 2018 Great Minds®. eureka-math.org

d. $354 \div 3$

e. $873 \div 4$

f. $591 \div 5$

Lesson 28: Represent and solve three-digit dividend division with divisors of 2, 3, 4, and 5 numerically.

© 2018 Great Minds®. eureka-math.org

EUREKA
MATH

g.   275 ÷ 3

h.   459 ÷ 5

i.   678 ÷ 4

Lesson 28:   Represent and solve three-digit dividend division with divisors
              of 2, 3, 4, and 5 numerically.

© 2018 Great Minds®. eureka-math.org

187

j.   955 ÷ 4

2. Zach filled 581 one-liter bottles with apple cider. He distributed the bottles to 4 stores. Each store
    received the same number of bottles. How many liter bottles did each of the stores receive?
    Were there any bottles left over? If so, how many?

Represent and solve three-digit dividend division with divisors
          of 2, 3, 4, and 5 numerically.

EUREKA
MATH

Name _____   Date _____

1. Divide. Check your work by multiplying. Draw disks on a place value chart as needed.

a. 776 ÷ 2	b. 596 ÷ 3

2. A carton of milk contains 128 ounces. Sara's son drinks 4 ounces of milk at each meal. How many 4-ounce servings will one carton of milk provide?

EUREKA MATH®   **Lesson 28:**   Represent and solve three-digit dividend division with divisors of 2, 3, 4, and 5 numerically.   189

© 2018 Great Minds®. eureka-math.org

Janet uses 4 feet of ribbon to decorate each pillow. The ribbon comes in 225-foot rolls. How many pillows will she be able to decorate with one roll of ribbon? Will there be any ribbon left over?

_____

_____

_____

_____

Read   **Draw**   **Write**

Lesson 29:   Represent numerically four-digit dividend division with divisors of 2, 3, 4, and 5, decomposing a remainder up to three times.

© 2018 Great Minds®. eureka-math.org

191

Name _____   Date _____

1. Divide, and then check using multiplication.

a. 1,672 ÷ 4

b. 1,578 ÷ 4

c. 6,948 ÷ 2

Lesson 29:   Represent numerically four-digit dividend division with divisors
of 2, 3, 4, and 5, decomposing a remainder up to three times.

193

d. 8,949 ÷ 4

e. 7,569 ÷ 2

f. 7,569 ÷ 3

Lesson 29:   Represent numerically four-digit dividend division with divisors
of 2, 3, 4, and 5, decomposing a remainder up to three times.

EUREKA
MATH

g. 7,955 ÷ 5

h. 7,574 ÷ 5

i. 7,469 ÷ 3

Lesson 29: Represent numerically four-digit dividend division with divisors of 2, 3, 4, and 5, decomposing a remainder up to three times.

195

j.   9,956 ÷ 4

2.   There are twice as many cows as goats on a farm. All the cows and goats have a total of 1,116 legs.
How many goats are there?

Lesson 29:   Represent numerically four-digit dividend division with divisors
of 2, 3, 4, and 5, decomposing a remainder up to three times.

© 2018 Great Minds®. eureka-math.org

EUREKA
MATH

Name _____   Date _____

1. Divide, and then check using multiplication.

a. 1,773 ÷ 3	b. 8,472 ÷ 5

2. The post office had an equal number of each of 4 types of stamps. There was a total of 1,784 stamps.
    How many of each type of stamp did the post office have?

Lesson 29:   Represent numerically four-digit dividend division with divisors
              of 2, 3, 4, and 5, decomposing a remainder up to three times.

197

© 2018 Great Minds®. eureka-math.org

The store wanted to put 1,455 bottles of juice into packs of 4. How many complete packs can they make? How many more bottles do they need to make another pack?

_____

_____

_____

_____

Read   **Draw**   **Write**

Lesson 30:   Solve division problems with a zero in the dividend or with a zero in the quotient.

199

© 2018 Great Minds®. eureka-math.org

Name _____ Date _____

Divide. Check your solutions by multiplying.

1. 204 ÷ 4

2. 704 ÷ 3

3. 627 ÷ 3

4. 407 ÷ 2

Lesson 30:   Solve division problems with a zero in the dividend or with a zero in the quotient.

© 2018 Great Minds®. eureka-math.org

201

5. 760 ÷ 4

6. 5,120 ÷ 4

7. 3,070 ÷ 5

8. 6,706 ÷ 5

Lesson 30: Solve division problems with a zero in the dividend or with a zero in the quotient.

© 2018 Great Minds®. eureka-math.org

EUREKA MATH

9.   8,313 ÷ 4

10.   9,008 ÷ 3

11. a.   Find the quotient and remainder for 3,131 ÷ 3.

b.   How could you change the digit in the ones place of the whole so that there would be no remainder? Explain how you determined your answer.

Lesson 30:   Solve division problems with a zero in the dividend or with a zero in the quotient.

203

© 2018 Great Minds®. eureka-math.org

Name _____   Date _____

Divide. Check your solutions by multiplying.

1. 380 ÷ 4

2. 7,040 ÷ 3

Lesson 30:   Solve division problems with a zero in the dividend or with a zero in
               the quotient.

205

© 2018 Great Minds®. eureka-math.org

1,624 shirts need to be sorted into 4 equal groups. How many shirts will be in each group?

_____

_____

_____

_____

Read   **Draw**   **Write**

Lesson 31:   Interpret division word problems as either *number* of groups *unknown*
or *group size unknown*.

© 2018 Great Minds®. eureka-math.org

207

Name _____ Date _____

Draw a tape diagram and solve. The first two tape diagrams have been drawn for you. Identify if the group size or the number of groups is unknown.

1. Monique needs exactly 4 plates on each table for the banquet. If she has 312 plates, how many tables is she able to prepare?

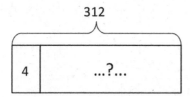

2. 2,365 books were donated to an elementary school. If 5 classrooms shared the books equally, how many books did each class receive?

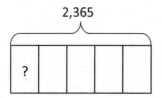

3. If 1,503 kilograms of rice was packed in sacks weighing 3 kilograms each, how many sacks were packed?

Lesson 31:   Interpret division word problems as either *number* of groups *unknown* or *group size unknown*.

© 2018 Great Minds®. eureka-math.org

209

4. Rita made 5 batches of cookies. There was a total of 2,400 cookies. If each batch contained the same number of cookies, how many cookies were in 4 batches?

5. Every day, Sarah drives the same distance to work and back home. If Sarah drove 1,005 miles in 5 days, how far did Sarah drive in 3 days?

Lesson 31:   Interpret division word problems as either *number* of groups *unknown* or *group size unknown*.

© 2018 Great Minds®. eureka-math.org

EUREKA
MATH

Name _____   Date _____

Solve the following problems. Draw tape diagrams to help you solve. Identify if the group size or the number of groups is unknown.

1. 572 cars were parked in a parking garage. The same number of cars was parked on each floor. If there were 4 floors, how many cars were parked on each floor?

2. 356 kilograms of flour were packed into sacks holding 2 kilograms each. How many sacks were packed?

Lesson 31:   Interpret division word problems as either *number* of groups *unknown*
or *group size unknown*.

211

Use the tape diagram to create a division word problem that solves for the unknown, the total number of threes in 4,194.

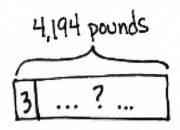

_____

_____

_____

_____

Read   Draw   Write

Lesson 32:   Interpret and find whole number quotients and remainders to solve
one-step division word problems with larger divisors of 6, 7, 8, and 9.

213

Name _____ Date _____

Solve the following problems. Draw tape diagrams to help you solve. If there is a remainder, shade in a small portion of the tape diagram to represent that portion of the whole.

1. A concert hall contains 8 sections of seats with the same number of seats in each section. If there are 248 seats, how many seats are in each section?

2. In one day, the bakery made 719 bagels. The bagels were divided into 9 equal shipments. A few bagels were left over and given to the baker. How many bagels did the baker get?

3. The sweet shop has 614 pieces of candy. They packed the candy into bags with 7 pieces in each bag. How many bags of candy did they fill? How many pieces of candy were left?

Lesson 32:   Interpret and find whole number quotients and remainders to solve one-step division word problems with larger divisors of 6, 7, 8, and 9.

© 2018 Great Minds®. eureka-math.org

215

4. There were 904 children signed up for the relay race. If there were 6 children on each team, how many teams were made? The remaining children served as referees. How many children served as referees?

5. 1,188 kilograms of rice are divided into 7 sacks. How many kilograms of rice are in 6 sacks of rice? How many kilograms of rice remain?

Lesson 32:   Interpret and find whole number quotients and remainders to solve one-step division word problems with larger divisors of 6, 7, 8, and 9.

© 2018 Great Minds®. eureka-math.org

Name _____   Date _____

Solve the following problems. Draw tape diagrams to help you solve. If there is a remainder, shade in a small portion of the tape diagram to represent that portion of the whole.

1. Mr. Foote needs exactly 6 folders for each fourth-grade student at Hoover Elementary School. If he bought 726 folders, to how many students can he supply folders?

2. Mrs. Terrance has a large bin of 236 crayons. She divides them equally among four containers. How many crayons does Mrs. Terrance have in each container?

Lesson 32:   Interpret and find whole number quotients and remainders to solve one-step division word problems with larger divisors of 6, 7, 8, and 9.

© 2018 Great Minds®. eureka-math.org

217

Write an equation to find the unknown length of each rectangle. Then, find the sum of the two unknown lengths.

3 m | 600 square m | 3 m | 72 square m

Read   Draw   Write

Lesson 33:   Explain the connection of the area model of division to the long
              division algorithm for three- and four-digit dividends.

© 2018 Great Minds®. eureka-math.org

219

Name _____   Date _____

1. Ursula solved the following division problem by drawing an area model.

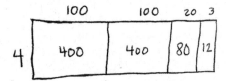

a. What division problem did she solve?

b. Show a number bond to represent Ursula's area model, and represent the total length using the distributive property.

_____

2. a. Solve 960 ÷ 4 using the area model. There is no remainder in this problem.

b. Draw a number bond and use the long division algorithm to record your work from Part (a).

Lesson 33:   Explain the connection of the area model of division to the long
             division algorithm for three- and four-digit dividends.

© 2018 Great Minds®. eureka-math.org

221

3. a. Draw an area model to solve 774 ÷ 3.

    b. Draw a number bond to represent this   c. Record your work using the long division
        problem.   algorithm.

---

4. a. Draw an area model to solve 1,584 ÷ 2.

    b. Draw a number bond to represent this   c. Record your work using the long division
        problem.   algorithm.

Lesson 33:   Explain the connection of the area model of division to the long
                                   division algorithm for three- and four-digit dividends.

                        © 2018 Great Minds®. eureka-math.org

EUREKA MATH

Name _____   Date _____

1. Anna solved the following division problem by drawing an area model.

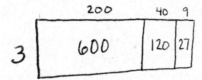

    a. What division problem did she solve?

    b. Show a number bond to represent Anna's area model, and represent the total length using the distributive property.

_____

2. a. Draw an area model to solve 1,368 ÷ 2.

    b. Draw a number bond to represent this problem.

    c. Record your work using the long division algorithm.

Lesson 33:   Explain the connection of the area model of division to the long division algorithm for three- and four-digit dividends.   **223**

© 2018 Great Minds®. eureka-math.org

Mr. Goggins planted 10 rows of beans, 10 rows of squash, 10 rows of tomatoes, and 10 rows of cucumbers in his garden. He put 22 plants in each row. Draw an area model, label each part, and then write an expression that represents the total number of plants in the garden.

_____

_____

_____

_____

Read   **Draw**   **Write**

Name _____   Date _____

1. Use the associative property to rewrite each expression. Solve using disks, and then complete the number sentences.

a.   30 × 24

= (_____ × 10) × 24

= _____ × (10 × 24)

= _____

hundreds	tens	ones

b.   40 × 43

= (4 × 10) × _____

= 4 × (10 × _____)

= _____

thousands	hundreds	tens	ones

c.   30 × 37

= (3 × _____) × _____

= 3 × (10 × _____)

= _____

thousands	hundreds	tens	ones

EUREKA MATH

Lesson 34:   Multiply two-digit multiples of 10 by two-digit numbers using a place value chart.

© 2018 Great Minds®. eureka-math.org

227

2. Use the associative property and place value disks to solve.
   a. 20×27

   b. 40×31

3. Use the associative property without place value disks to solve.
   a. 40×34 b. 50×43

4. Use the distributive property to solve the following problems. Distribute the second factor.
   a. 40×34 b. 60×25

EUREKA
MATH®

Name _____ Date _____

1. Use the associative property to rewrite each expression. Solve using disks, and then complete the number sentences.

   20×41

   ____ × ____ × ____ = ____

hundreds	tens	ones

2. Distribute 32 as 30 + 2 and solve.

   60×32

EUREKA
MATH

Lesson 34: Multiply two-digit multiples of 10 by two-digit numbers using a place value chart.

229

For 30 days out of one month, Katie exercised for 25 minutes a day. What is the total number of minutes that Katie exercised? Solve using the place value chart.

thousands	hundreds	tens	ones

_____

_____

_____

_____

Read   **Draw**   **Write**

Lesson 35:   Multiply two-digit multiples of 10 by two-digit numbers using the area model.

© 2018 Great Minds®. eureka-math.org

231

Name _____   Date _____

Use an area model to represent the following expressions. Then, record the partial products and solve.

1. 20 × 22

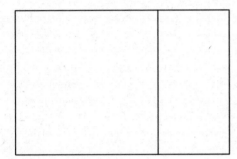

```
    2 2
  ×  2 0
  _____

+ _____
  ━━━━━━━
```

2. 50 × 41

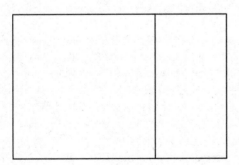

```
    4 1
  ×  5 0
  _____

+ _____
  ━━━━━━━
```

3. 60 × 73

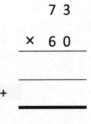

```
    7 3
  ×  6 0
  _____

+ _____
  ━━━━━━━
```

Draw an area model to represent the following expressions. Then, record the partial products vertically and solve.

4. 80 × 32

5. 70 × 54

Visualize the area model, and solve the following expressions numerically.

6. 30 × 68

7. 60 × 34

8. 40 × 55

9. 80 × 55

Lesson 35: Multiply two-digit multiples of 10 by two-digit numbers using the area model.

© 2018 Great Minds®. eureka-math.org

EUREKA MATH

Name _____   Date _____

Use an area model to represent the following expressions. Then, record the partial products and solve.

1. 30 × 93

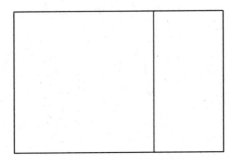

```
      9 3
  ×   3 0

  _____
+
  _____
```

2. 40 × 76

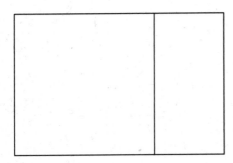

```
      7 6
  ×   4 0

  _____
+
  _____
```

EUREKA MATH

Lesson 35:   Multiply two-digit multiples of 10 by two-digit numbers using the area model.

© 2018 Great Minds®. eureka-math.org

235

Mr. Goggins set up 30 rows of chairs in the gymnasium. If each row had 35 chairs, how many chairs did Mr. Goggins set up? Draw an area model to represent and to help solve this problem.

_____

_____

_____

_____

Read   **Draw**   **Write**

Lesson 36:   Multiply two-digit by two-digit numbers using four partial products.

237

© 2018 Great Minds®. eureka-math.org

Name _____   Date _____

1. a. In each of the two models pictured below, write the expressions that determine the area of each of the four smaller rectangles.

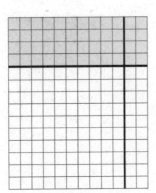 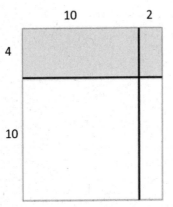

b. Using the distributive property, rewrite the area of the large rectangle as the sum of the areas of the four smaller rectangles. Express first in number form, and then read in unit form.

14 × 12 = (4 × _____) + (4 × _____) + (10 × _____) + (10 × _____)

2. Use an area model to represent the following expression. Record the partial products and solve.

14 × 22

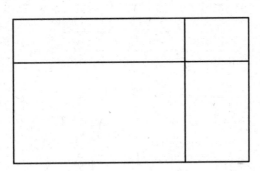

```
      2 2
  ×   1 4
  _____
  _____
  _____
  _____
+ _____
```

Lesson 36:   Multiply two-digit by two-digit numbers using four partial products.

239

EUREKA
MATH

© 2018 Great Minds®. eureka-math.org

Draw an area model to represent the following expressions. Record the partial products vertically and solve.

3. 25 × 32

4. 35 × 42

Visualize the area model and solve the following numerically using four partial products. (You may sketch an area model if it helps.)

5. 42 × 11

6. 46 × 11

Lesson 36:   Multiply two-digit by two-digit numbers using four partial products.

© 2018 Great Minds®. eureka-math.org

EUREKA
MATH

Name _____ Date _____

Record the partial products to solve.

Draw an area model first to support your work, or draw the area model last to check your work.

1. 26 × 43

2. 17 × 55

Sylvie's teacher challenged the class to draw an area model to represent the expression 24 × 56 and then to solve using partial products. Sylvie solved the expression as seen below. Is her answer correct? Why or why not?

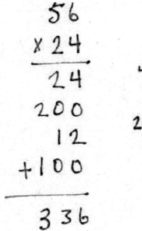

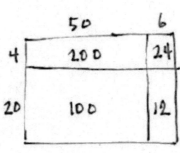

_____

_____

_____

_____

Read   **Draw**   **Write**

Lesson 37:   Transition from four partial products to the standard algorithm for two-digit by two-digit multiplication.

243

EUREKA
MATH®

© 2018 Great Minds®. eureka-math.org

Name _____ Date_____

1. Solve 14 × 12 using 4 partial products and 2 partial products. Remember to think in terms of units as you solve. Write an expression to find the area of each smaller rectangle in the area model.

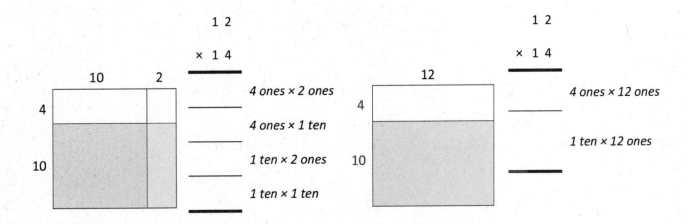

```
  1 2
× 1 4
─────
_____   4 ones × 2 ones

_____   4 ones × 1 ten

_____   1 ten × 2 ones

_____   1 ten × 1 ten
─────
```

```
  1 2
× 1 4
─────
_____   4 ones × 12 ones

_____   1 ten × 12 ones
─────
```

2. Solve 32 × 43 using 4 partial products and 2 partial products. Match each partial product to its area on the models. Remember to think in terms of units as you solve.

```
  4 3
× 3 2
─────
_____   2 ones × 3 ones

_____   2 ones × 4 tens

_____   3 tens × 3 ones

_____   3 tens × 4 tens
─────
```

```
  4 3
× 3 2
─────
_____   2 ones × 43 ones

_____   3 tens × 43 ones
─────
```

EUREKA MATH

Lesson 37: Transition from four partial products to the standard algorithm for two-digit by two-digit multiplication.

245

© 2018 Great Minds®. eureka-math.org

3. Solve 57 × 15 using 2 partial products. Match each partial product to its rectangle on the area model.

4. Solve the following using 2 partial products. Visualize the area model to help you.

a.
```
      2 5
   ×  4 6
   _____

              _____ × _____

   _____

              _____ × _____

   _____
```

b.
```
      1 8
   ×  6 2
   _____

              _____ × _____

   _____

              _____ × _____

   _____
```

c.
```
      3 9
   ×  4 6
   _____
```

d.
```
      7 8
   ×  2 3
   _____
```

Lesson 37:   Transition from four partial products to the standard algorithm for two-digit by two-digit multiplication.

© 2018 Great Minds®. eureka-math.org

EUREKA
MATH

Name _____   Date _____

1. Solve 43 × 22 using 4 partial products and 2 partial products. Remember to think in terms of units as you solve. Write an expression to find the area of each smaller rectangle in the area model.

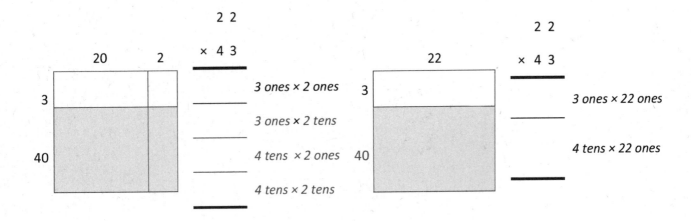

2. Solve the following using 2 partial products.

                6 4
      ×   1 5
     ——————

5 ones × 64 ones

1 ten × 64 ones

EUREKA MATH

Lesson 37:   Transition from four partial products to the standard algorithm for two-digit by two-digit multiplication.

© 2018 Great Minds®. eureka-math.org

247

Sandy's garden has 42 plants in each row. She has 2 rows of yellow corn and 20 rows of white corn. Draw an area model (representing two partial products) to show how much yellow corn and white corn has been planted in the garden.

_____

_____

_____

Read   **Draw**   **Write**

Lesson 38:   Transition from four partial products to the standard algorithm for
two-digit by two-digit multiplication.

249

EUREKA
MATH

Name _____   Date _____

1. Express 23 × 54 as two partial products using the distributive property. Solve.

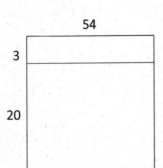

23 × 54 = (___ fifty-fours) + (___ fifty-fours)

```
    5 4
×   2 3
_____

_____   3 × _____

           20 × _____
_____
```

2. Express 46 × 54 as two partial products using the distributive property. Solve.

46 × 54 = (___ fifty-fours) + (___ fifty-fours)

```
    5 4
×   4 6
_____

_____   _____ × _____

           _____ × _____
_____
```

3. Express 55 × 47 as two partial products using the distributive property. Solve.

55 × 47 = (____ × _____) + (_____ × _____)

```
    4 7
×   5 5
_____

_____   _____ × _____

           _____ × _____
_____
```

EUREKA
MATH

Lesson 38:   Transition from four partial products to the standard algorithm for
two-digit by two-digit multiplication.

251

4. Solve the following using 2 partial products.

```
        5 8
    ×   4 5
    ─────────
    _____   ____ × ____
    ─────────
    ═════════   ____ × ____
```

5. Solve using the multiplication algorithm.

```
        8 2
    ×   5 5
    ─────────
    _____   _____ × _____
    ─────────
    ═════════   _____ × _____
```

6. 53×63 7. 84×73

Lesson 38:   Transition from four partial products to the standard algorithm for
                    two-digit by two-digit multiplication.

EUREKA
MATH

Name _____   Date _____

Solve using the multiplication algorithm.

1.

$$
\begin{array}{r}
7\ 2 \\
\times\quad 4\ 3 \\
\hline
\end{array}
$$

_____ × _____

_____ × _____

2. 35×53

Lesson 38: Transition from four partial products to the standard algorithm for two-digit by two-digit multiplication.

© 2018 Great Minds®. eureka-math.org

253

Credits

Great Minds® has made every effort to obtain permission for the reprinting of all copyrighted material. If any owner of copyrighted material is not acknowledged herein, please contact Great Minds for proper acknowledgment in all future editions and reprints of this module.

9-23-03

SMOKE
SCREEN

SMOKE
SCREEN

KYLE MILLS

G. P. PUTNAM'S SONS   NEW YORK

This book is a work of fiction. Names, characters, places, and incidents either
are the product of the author's imagination or are used fictitiously,
and any resemblance to actual persons, living or dead, business
establishments, events, or locales is entirely coincidental.

G. P. Putnam's Sons
Publishers Since 1838
a member of
Penguin Group (USA) Inc.
375 Hudson Street
New York, NY 10014

Copyright © 2003 by Kyle Mills
All rights reserved. This book, or parts thereof, may not
be reproduced in any form without permission.
Published simultaneously in Canada

Library of Congress Cataloging-in-Publication Data

Mills, Kyle, date.
Smoke screen / Kyle Mills.
p. cm.
ISBN 0-399-15098-6
1. Tobacco industry—Fiction.   2. Corporate Culture—Fiction.   I. Title.
PS3563.I42322S56   2003   2003046746
813'.54—dc21

Printed in the United States of America
1   3   5   7   9   10   8   6   4   2

This book is printed on acid-free paper. ∞

BOOK DESIGN BY MEIGHAN CAVANAUGH

ACKNOWLEDGMENTS

Thanks to Chris Huther and Francie Makris for their help in researching a jurisdictional issue that was well over my head. Any legal errors are mine. Also, apologies to Chris for making fun of him in the text. I couldn't resist, and I suspect that he and I are the only people in the world who will get the joke.

As always, thanks to my wife, Kim, and my parents, Darrell and Elaine, for their comments throughout the writing process. Also to my editor, Rob McMahon, who was instrumental in helping me tighten and focus the book.

Laura Liner for her help on *Sphere of Influence* and for putting me in touch with the right people on this book.

Special thanks to D., without whom I don't think I could have pulled this off. Her willingness to thoroughly answer endless questions about the tobacco industry and her patience when I couldn't understand those answers went beyond generous.

More persons are, on the whole, humbugged by believing nothing than by believing too much.

P. T. BARNUM

SMOKE
SCREEN

MY DOG SUDDENLY DECIDED TO ABANDON HER ROLE AS OTTOMAN and sprang to her feet with enough force to almost flip me backward in my chair. She tensed and faced the television, concentrating with puppy-like enthusiasm on the man speaking through it. I watched too, but a little less passionately.

"The new surgeon general's report is very simply above reproach! We're talking about research techniques that have been in the process of refinement for over half a century now. We're talking about the most up-to-date scientific and statistical methods. We're talking about some of the top minds in the field . . ."

"Wait for it . . . ," I said as a playful growl escaped my aging Great Pyrenees. She crept forward a few inches but then stopped.

"The fact that Big Tobacco would even bother to send someone here to try to refute these findings is absurd and insulting. But not exactly surprising, I suppose."

The man speaking was Angus Scalia, the tobacco industry's most rabid critic and a fairly difficult person to describe: Picture a sixty-year-old, four-hundred-pound John Lennon suffering from male pattern baldness and trying to hide it with a ponytail comb-over. Dress him in clothing purchased exclusively at Texas outlet malls and you've got a pretty close approximation.

"There is no longer any doubt that cigarette smoking is the greatest threat to public health since the Black Death decimated Europe in the fourteenth century."

"Good one," I said aloud, unable to fend off the mental image of those old woodcuts depicting a personified Death hovering over crying children as they watched their fathers' sore-covered bodies stacked onto a wheelbarrow.

While I'd never admit it publicly, I couldn't help admiring Scalia's unwavering conviction. For him, it wasn't about politics or getting his face on TV. This guy really, truly, believed. He was a man who could actually make a coherent argument that all things evil in the modern world were in some way connected to Big Tobacco. That kind of single-mindedness had a strange purity to it that I found fascinating.

With Scalia slightly out of breath and obviously in need of a rest, the camera cut to a gray-haired man in his early fifties nodding with a convincing facsimile of sincerity. I couldn't immediately remember his name—something foreign sounding that started with one of those letters that didn't get much play in the English language: *Q? X? V?*

Viasanto. That was it. Craig Viasanto.

"Let me tell you a story, Mr. Scalia . . . ," he started.

My dog lunged, but I saw it coming and managed to stop her by wedging a sock-covered foot beneath her collar.

"Many years ago, R. J. Reynolds decided he wanted to expand out of the chewing-tobacco market and into cigarettes. But he didn't want to get involved in selling something that might be dangerous, so he had three independent labs do studies on the health ramifications of smoking. All three responded that there was no danger. None at all. The point I'm trying to make is that we should be careful about putting too much faith in 'scientific studies.' It used to be that vitamin C was the great cure-all, but now the medical community is suggesting that it may cause genetic damage. And remember when hormone replacement was the best thing ever to happen to postmenopausal women? Now we're hearing about thirty percent increases in heart attack and breast cancer rates."

Calmly earnest, impeccably groomed, and carefully studied, Craig Viasanto couldn't have been a more striking counterpoint to Angus Scalia. Based on his Botoxlike ability to say absolutely anything with a straight face, he'd recently been made the tobacco industry's top spokesman. I, of course, had protested his promotion by firing off a five-page memo outlining the benefits of hiring Pam Anderson as the industry's primary spokesperson. I, for one, was prepared to believe anything Ms. Anderson told me.

Surprisingly, I never received a response to my eminently reasonable and exhaustively researched proposal.

"It would be naïve of us to believe that politics doesn't have a profound effect on science," Viasanto continued. "I could name a thousand examples, going as far back as the Catholic Church's insistence that the sun revolved around the Earth."

I gave him a little golf clap for that one, but my dog was still straining at her collar and I had to keep hold of the chair to avoid being dragged onto the floor. The reference was subtle, but clearly alluded to a priest who just yesterday had admitted to molesting no fewer than thirty young boys in his care. The media was all over it, and Viasanto was obviously not above using this revelation to channel the public's ire in another direction.

"It's the height of fashion right now to blame the tobacco industry for everything from lung cancer to the budget deficit. We're quickly becoming the twenty-first-century version of McCarthy's Communists."

He'd gone a little overboard there and couldn't quite sell it. You had to applaud the effort, though.

"Let me tell *you* a story, Mr. Viasanto," Scalia said as the camera refocused on him. Rage had contorted his thick face to the point that it seemed to be in the process of swallowing the tiny round glasses perched on his nose.

"After World War One, a soldier who had been provided cigarettes— a previously obscure product—by the U.S. Army, died of lung cancer. Before performing an autopsy, one of the foremost doctors of the time called all his colleagues and invited them. He said, and I quote, 'You'll never see another one of these again.' That's how rare this disease used to be."

"Whoops," I said aloud.

"A perfect example of what I'm talking about," Viasanto said, immediately pouncing on Scalia's rare misstep. "No one ever mentions the explosion of industrial pollution and the shift in the population from rural to urban areas when we talk about lung ailments."

Scalia knew he'd walked into that one, and his bolo tie suddenly seemed dangerously tight. "This is completely ridiculous! You know smoking is dangerous as well as I do! Here's another quote for you: 'We agree that smoking is addictive and causes disease in smokers.' One of your own vice presidents said that!"

Before Viasanto could respond, the camera cut to the show's host and he announced a brief commercial break. My dog relaxed a little.

"Why do I watch these things?" I said aloud.

All they did was make me question the value of the human race and, more particularly, my own precarious position in that race. Besides, it was six-thirty and I knew for a fact that there was a rerun of *Three's Company* on channel 283.

The news program flashed back on and I pretended to ignore it, though for whose benefit I'm not sure. Mine, I suppose.

"Mr. Scalia," the host resumed, "you've been quoted as saying that the class-action suit in Montana is the beginning of the end for the tobacco industry. But is that realistic? We're talking about an industry that helped found this country and now accounts for almost one percent of American employment."

A little background is necessary:

A $250-billion class-action lawsuit had been filed in Montana alleging (surprise) that the tobacco industry was selling a deadly product, that it was encouraging people to use that product, and that the dangers of the habit had been, and continued to be, played down and lied about. All completely true. In fact, it wouldn't be hard to mistake the complaint for a tobacco company's mission statement.

To be fair, though, it's a little more complicated than that. There are the government warnings, the highly subjective concepts of personal responsibility and freedom, and the question of whether anyone reasonably believes that smoking isn't killing them.

As a practical matter, though, these valid issues were sometimes completely irrelevant to the tobacco industry's courtroom performance. What it tended to come down to was just how pissed off the jury was. And generally they're pretty pissed off, which is why Big Tobacco does a lot better in front of appellate judges than it does in front of real people.

So what was going on in Montana that had everyone so worked up? I mean, the jury would almost certainly award the plaintiffs the 250 billion and then the industry would unleash that clever and well-coiffed legal hoard it kept penned up in its basement and the judgment would be overturned on appeal. It was the way of the world. Right?

In this case, maybe not.

Montana doesn't have a law that sets a maximum on the amount of an appeals bond. That means if Big Tobacco were to lose this particular case, the companies would have to come up with the entire judgment amount in order to file an appeal. And contrary to popular opinion, they just didn't have that kind of money sitting around.

The bottom line? If a Montana jury decided to hold the tobacco industry responsible for rain in July and then awarded more money than the industry could bond off, that judgment would stick. And the entire house of cards would collapse.

"I absolutely think it's realistic," Scalia said. "In fact, I think it's likely. Big Tobacco has lost its ability to control and subvert the information available to the American people. These companies have finally been exposed as the despicable liars they are. Now, politicians know that if they take money from Tobacco and become apologists for these murderers, their constituents see them for what they are: whores. Mark my words, Tom—we'll see the complete implosion of the industry in the next ten years."

The host nodded thoughtfully and turned back to Viasanto. "Do you have any comment on that?"

For the first time I could remember, the industry's savant of a mouthpiece looked a little lost. "I'm sorry, Tom. I'm not at liberty to comment on ongoing litigation."

My dog strained forward again when Viasanto spoke and I finally let my foot slide from beneath her collar.

"Go get him, girl!"

She barked joyfully and leaped forward, covering my TV screen with paw prints and saliva. The less than vicious attack lasted only a few seconds before she exhausted herself and collapsed on the carpet for a well-deserved nap.

I suddenly realized that I was in desperate need of a drink. In fact, I was in need of a bunch of drinks—more than my fridge could comfortably provide. I pushed myself to my feet, turned off the TV, and started for the front door.

"I'll be back in a few hours, Nicotine. Can you hold down the fort for a while?"

She snorted and rolled farther onto her back but didn't open her eyes.

1 "So do you *have* to be naked to use the foosball table?"

At least that's what I think she said. The house's half-million-dollar sound system was being pushed to its limit by one of those repetitive, mechanical-sounding drones people slightly younger than me liked to listen to. I focused on her mouth as she spoke, trying to read her lips through the smoke and chaotic lighting, but found myself concentrating on their plump perfection instead.

I managed to turn slightly, bumping someone behind me and sending most of his beer down my back. It felt pretty good, so I shrugged off his apology and looked out across a dance floor so full that it made nonvertical motion completely impossible. On the downbeats, I could just make out some bare skin over the top of the pogoing crowd.

"I'm not really sure," I shouted loud enough for her to hear but not so loud as to shower her with spit. "I think it's more of a guideline."

She mulled that over for a moment. "Why?"

Now, that was a question that demanded an answer that was probably too long and complicated to get across in the current setting.

A hundred years ago, the house we were standing in had been the calculatedly imposing home of a wealthy plantation owner—a man who still sat, white suit and all, in an old daguerreotype above the toilet in one of the bathrooms. In its heyday, the house had been filled with European furniture, South American silver, and Chinese silk—all carefully maintained by a staff of ex-slaves who would have been well on their way to realizing that freedom was a more ethereal concept than they'd originally thought. Parties, frequent and lavish, would have been carefully planned to highlight the breeding and superiority of guests wandering stiffly through it and to nudge upward the social standing of the hosts.

All that was gone now, replaced by the previously mentioned sound system, an elaborate bar constructed out of an old VW microbus, no less than five big-screen TVs, concert lighting, and an undetermined number of sweating, occasionally naked, twenty-somethings. Outside, the once-stately gardens had been replaced with a twenty-person hot tub, a pool in the inexplicable shape of a star, and an inoperable crane that would soon be coaxing gravity-assisted projectile vomiting from aspiring bungee jumpers.

I shrugged as the girl and I sidestepped away from the expanding mass on the dance floor. I hadn't caught her name, or maybe I had and just couldn't remember. "I don't know. Tradition, I guess."

She tilted her head, causing her nose ring to flash hypnotically as she tried to decide if I was making fun of her.

I'd been trying to peg her age for the last fifteen minutes, but she was one of those people who seemed to gain and lose years with every change of expression. My current best guess was that she was a few years younger than me. Say, twenty-eight.

"You were telling me about your trip," I yelled, trying to divert the current flow of conversation, which would inevitably lead to questions about the infamous owner of the house—a subject to be avoided at all costs.

"After I left MIT, I did some traveling—you know, just put my bike on a plane went where it was cheap. I started in Europe . . . Have you ever been to Prague?"

I shook my head.

"Beautiful city—and you can hang out there for next to nothing. I rode across the Czech Republic—"

"By yourself?"

"Yeah. I was supposed to go with a bunch of friends, but they all got jobs right out of school and backed out on me. It was better, though, you know? I was kind of forced to dive into the local culture. The people were great—they took me in, let me stay at their houses . . . I even slept in barns sometimes with the cows."

I grinned, probably stupidly. "Really? Cows?"

"Hey, don't laugh. After a day of hammering your bike through the rain, you'd be psyched to curl up with a cow. They generate a lot of heat."

"Yeah, I guess I can see how they would . . ."

"So anyway, then I hopped a train and headed up to Scandinavia. Ever been to Copenhagen?"

A guy in a Superman costume climbed up on a pool table and dove into the crowd on the dance floor. I watched him surf along on a cushion of up-turned hands and then get ejected onto the floor a few feet from us.

"Denmark? Huh uh."

"Nice place. Nice people. And everyone speaks English, which was a good change. But man, it's expensive. I only stayed a few days before I headed south again. I spent another month just cruising around and then shipped my bike back to my folks and headed for Asia. Ever been to Thailand?"

"Never."

"Really exotic. You should try to make a trip. Great food, super cheap."

"Yeah, one of these days. Hard to find the time, you know?"

"Yeah, they work you pretty hard here," she said, leaning in a little closer to my ear. "I've only been at the company for six months. Kind of hard to make the adjustment from just screwing around all the time. What division do you work in? I haven't seen you around . . . Seems like I would have remelmbered if I had."

I managed not to wince visibly when the cramped muscles at the base of my neck spasmed. Had that been an expression of interest?

The girl was beautiful, intelligent, looked good in a nose ring, told jokes about Tolstoy that were actually kind of funny, and was talking to me instead of one of the hundred other guys patrolling the area. I wasn't good under this kind of pressure.

She smiled, displaying a set of teeth that were well worth whatever her parents had paid for them. "Yeah, I definitely would have remembered."

First impressions of me were as varied as the personalities that formed them. I'm just a bit over six foot four, with thick shoulders and a narrow, well-defined waist that hasn't yet succumbed to my more-or-less sedentary lifestyle. It was a physique that provoked lust, envy, and intimidation, among other things.

My light blond hair, sun-deprived skin, and teeth that were overly white despite my best efforts to yellow them invited comparisons to angels and Nazis in roughly equal proportions.

The bad habit I had of holding conversations and maintaining relationships while almost never making eye contact made a few people paint me as shy, but most complain about my arrogance.

"I, uh, don't work for the company. I'm just friends with the owner," I stammered and then immediately cursed myself for being so stupid.

"Darius? You mean you know Darius? You're friends with him? No way!"

"Uh yeah. I guess I know him okay," I said, trying to backpedal.

Actually, we'd been best friends since the fifth grade. In fact, we'd been inseparable enough that Darius had followed me to Chico State, despite having nearly every Ivy League dean in the country willing to prostitute himself in surprisingly degrading ways to lure him to one of America's more esteemed learning institutions.

Honestly, I'd taken his willingness to drive me to California as just another excuse for a multistate crime spree. When we arrived—literally just hours ahead of the authorities—he'd disappeared, leaving me to wrestle my stuff out of the car and up the stairs to my dorm room. When he'd finally reappeared, he'd lost his shirt and socks, but gained a blanket, an alarm clock, and a full academic scholarship.

Five years later, with a total of thirty credit hours and a D average, security had escorted him from school grounds for the last time. So what did he do then? What any self-respecting college dropout would do: Stumbled drunkenly to the bank, withdrew the money he'd earned by skipping class and doing cutting-edge programming jobs, and started a computer game company. Now when God needed a loan, He called Darius.

"Is he here?" The Girl said.

I shrugged, and surprisingly the conversation moved back in my direction.

"So if you don't work here, what do you do?"

"I'm a, uh, trustafarian."

She scrunched up her nose in away that was irresistibly cute. "A trustafarian? What's a trustafarian?"

I considered my answer, sipping what was left of my warm vodka tonic to cover up the fact that I was stalling. As with all things, the answer to her question was a matter of perspective . . .

Trustafarian 'trəst-ə-'far-ē-an n. 1: A person who inherited his or her money in the form of a trust, which pays said funds out in installments. 2: One who lives off the hard work and resourcefulness of dead relatives who weren't smart enough to blow their money on women and booze. 3: A person who lives well but contributes nothing of value to society. 4: A lazy, good-for-nothing leech on society who will never get anywhere in life with that attitude . . .

"So you don't do anything? Nothing at all?" She waved her hand around. "You go to parties?"

I sighed inaudibly at her description of the European-Arab ideal. In those more civilized societies, having old money and never working a day in your life made everyone think you were better than they were. It was so much more complicated in America.

"I work in the family business," I said.

She nodded, actually looking interested as opposed to just trying to wrangle the conversation back around to Darius the Great. When she pulled a pack of cigarettes from her pocket, that expression of interest faded into one of apology.

"Do you mind?" she said, patting her pockets for a lighter. I pulled mine out and lit the cigarette for her.

"You want one?" she asked.

They weren't my brand but I took one anyway, lighting it and taking a characteristically shallow drag.

"You were telling me about your family's business," she shouted.

"Was I?"

"Well, you were about to."

Oddly, I am not a liar by nature. But I do occasionally succumb, if the lies are white and ultimately temporary.

"We invented those little felt things that go on the bottoms of chairs to keep them from scratching your floor."

"Felt?"

"Not felt per se—just the application of felt to the legs of furniture."

In my experience, it is virtually impossible to talk about felt for more than three minutes. The current record was about two and a half.

"Who'd have thought there was a ton of money in that?"

"There's not," I said. "Honestly, it's not a very good trust. But every little bit—"

I fell silent when I saw the well-defined edge of the crowd on the dance floor turn liquid and a small wave form as people briefly retreated toward the middle and then moved back out to the edge. I couldn't be sure what was causing the strange disturbance, but I had a pretty good guess.

"Every little bit what?" I heard The Girl say. I moved closer to her and turned toward the wall, trying to hide and block her from view at the same time. It was too late, though. The volume of the music started to decline at an almost imperceptible rate and was soon down to a level that would allow communication at a slightly more dignified volume.

"Programmer. Tina. New, right?"

Darius tended to talk like that. Single. Words. Particular order? None. I stepped reluctantly aside and he moved in, casually smoothing the silky brown hair hanging loosely around his shoulders. For some reason—sheer willpower probably—the blue-tinted, rectangular glasses he loved so much hadn't been fogged by the wet heat surrounding us and he peered over them at the girl whose name I now knew.

"Um, yeah, right . . . ," Tina said, compulsively twisting her hair around her index finger. "How did you know?"

Darius put his hands out in front of him and wiggled his fingers like a proud magician. The music faded another subtle notch. "I own the company. I know all. I've never seen you here. Is this your first party?"

She nodded.

These little get-togethers were invitation-only, and the guest list was one of the few things Darius didn't delegate these days. He knew damn well that this was her first time here and had undoubtedly not invited her until now because he was backlogged with all the other beautiful young employees his personnel department had amassed.

"Are you having a good time?"

"Really good!" she said—not quite gauging the new sonic environment correctly. The answer, or more precisely, the overly loud and nervous delivery of it, seemed to please Darius.

"I was . . . I was just talking to your friend," Tina said. "He was telling me about his job." She turned to me, a little wider-eyed than she'd been earlier, willing me to speak and make sure she didn't make a fool of herself in front of DariSoft's President, CEO, and Svengali.

Darius's head swiveled in my direction, but his body remained squared to Tina's. It was hard for me to see him the way Tina did. I'd become accustomed to how hot he burned, but I'd experienced his initial effect on people enough times to understand a little of what she was feeling. I pretended to be jostled by the throng behind me and stepped into Tina again, shoving Darius aside.

"Man, it's crazy in here," I said. "Why don't we catch up with you outside, Darius. I know how much you hate crowds."

He looked at me the way my father used to when I back-talked him. "You trying to get rid of me, Trevor?" We locked eyes for a moment, and I backed up a step.

"So he was telling you about his job, huh . . . ," Darius said to Tina, glancing over at me as he spoke. Thoughts of twisting his head off like a bottle cap flew across my mind, but instead I just stood there.

"Let me guess. Porta Potties?"

I stared down at my beer-soaked feet, but I could feel Tina's eyes on me. I wasn't sure whether that was out of growing suspicion or the fact that gazing upon Darius with the naked eye was difficult for some people. I took another shallow drag on the cigarette she'd given me.

"No?" I heard Darius say. "Hmm. Electric nose hair clippers?"

No audible response from Tina, but I guessed that she was shaking her head.

"We're not back to those little felt things you put on chairs, are we?"

Still nothing from Tina. Nodding, I suppose.

Darius slung an arm around my stooped shoulders and laughed, then took a dramatic pull from the bottle of Jack Daniel's that he was rarely without at these parties.

"So your family didn't invent those felt things?" I heard Tina say.

You'd think I'd have devised a clever way out of these types of situations, but for some reason I never had.

"Are you kidding?" Darius said, giving me a friendly squeeze. "Trevor's family practically invented the tobacco industry in this country."

And there it was.

I dared a quick peek and, as expected, she was looking down at the cigarette in her hand.

The next few seconds would be critical. In my experience, nine out of ten young, healthy smokers halfheartedly supported the industry that provided them with such a pleasurable, relaxing, weight-controlling, image-enhancing product. The other one acted as though they'd met their future murderer.

"Another liar from the tobacco industry," Tina said, dropping her cigarette into what was left of my drink. Darius and I watched her push her way through the crowd and finally disappear through the doors that led to the pool.

"Bitch doesn't have much of a sense of humor, does she?" He had an expression of what could only be described as anesthetized pain on his face. I frowned.

"What's that look for?" he said, putting his bottle to my cup and filling it, despite the fact that it still contained Tina's cigarette. I kept my disapproving glare aimed an him, and he glared back.

"Oh, quit pouting, Trevor. You're acting like a big baby. She was going to find out anyway."

He'd never looked away first in all the years I'd known him. I finally turned slightly and pretended to scan the crowd.

"Why don't you just come to work for me, Trev? Chicks dig computer programmers." He took another belt from his bottle and his arm across my shoulders became less an act of friendship than a means to stay upright.

"No they don't."

"Better than tobacco industry bigwigs."

"I don't know anything about computers."

"You could bring us coffee."

He laughed until he doubled over and started coughing. Being the helpful soul I am, I slapped him on the back hard enough to almost cause his knees to buckle beneath him. He wisely moved to a safer distance, smoothed his hair one last time, and then sped off after Tina.

2 "Someone . . . Kill me."

I think I spoke the words aloud, and I think I was serious, but wouldn't swear to it. The gushing buzz in my ears was keeping perfect rhythm with the peaks and valleys of pain in my head, and the North Carolina sun flooding through the window felt burning hot. I tried to roll away but found that complete stillness was the only thing keeping me from vomiting. So I was stuck. Trapped by my own stupidity.

The events that had left me in this sorry state came back slowly: I remembered the crowd, the heat, the girl. I remembered beer and vodka and whisky. I also had a hazy recollection of trying to set Darius on fire with a cigarette and a glass of tequila. I'd failed miserably, and a hastily assembled jury of revelers sentenced me to three shots of the surprisingly difficult to ignite fluid. After that there was nothing. Zip.

How had I gotten home?

Surely I hadn't driven. Had someone given me a ride? A brief surge of adrenaline tried to crowbar my skull apart.

The Girl.

Maybe she'd decided to ignore the tobacco hysteria being whipped up by the media and consider the years of cool menthol goodness I was helping to provide her. It took me a few minutes to retrieve her name, and by then I had convinced myself that Tina was right there in bed next to me.

I could almost see that nose ring of hers glimmering in the sun as she breathed. That had to be it. Tina'd driven me.

Despite my certainty, I didn't move or open my eyes. I told myself it was because of the nausea, but the truth was I didn't want to disturb my fantasy. It wasn't often that I found myself lying only inches from the perfect woman—the solution to my loneliness, boredom, doubt . . . If I wanted to, I could just reach out and . . .

The barely audible clicking of claws on wood started somewhere in the house, becoming louder and more staccato as it approached. I ignored the sound until the mattress suddenly dipped and a wet, pizza-scented tongue slapped me on the side of the head. The combination of the motion and Italian dog breath turned out to be a little more than I could handle. My stomach pitched into my throat and I jerked myself into a sitting position, hoping that gravity would put my insides back where they belonged. I swallowed hard a few times and managed, barely, to keep it together. When the room stopped moving, I dared a glance over at the other side of the bed to confirm what I really already knew. Tina wasn't there.

"Look what you did," I croaked. "You scared her off."

Nicotine leaned farther onto the bed, sniffed at the jacket I was still wearing, and grimaced as well as evolution would allow.

"Yeah, well you don't smell that good either," I said, a little too defensively. "You got into the fridge again, didn't you?"

She slunk away, heading for the bathroom to escape the stench of cigarettes, sweat, and alcohol. I slid one foot to the floor but then thought better of it and eased myself back onto my pillow and back into unconsciousness.

It was almost eight when my feet finally hit the floor for real. I grabbed an open pack of cigarettes off the nightstand, but couldn't bring myself to light one. I promised myself I'd double up later.

Nicotine was curled up in front of the shower doing her impression of a big white bath mat, so I had to lean over her to turn the water on. Lukewarm, just the way she liked it.

"Well? What are you waiting for?" I said, pushing the glass door open and starting to strip. "Go on."

Now, I don't normally shower with my dog—primarily because the stall isn't that big and sharing it with a soggy 150-pound canine left very little room to maneuver. But she loved the shower almost as much as she hated it when I left her alone on Sunday night, which was traditionally our popcorn and movie time. And for a dog, she could really lay on the guilt.

"You'd have liked her, Nicky," I said, during the brief scuffle in which I gained control of the showerhead. "She went to MIT and traveled all over the world. About five-six . . . Almost-black hair . . ."

Nicotine was concentrating on trying to capture the bubbles generated by my shampoo, leaving me with the distinct feeling that she wasn't listening.

The truth was that she had nothing but distaste for the occasional women I brought home—mostly your typical debs and southern belles. Not my first choice either, but they tended to be fairly easy to chase down and catch.

The really interesting women—the ones who told jokes about Tolstoy—were so much harder to come by. Particularly with the media bored stiff by a quietly whimpering economy, a fizzled terrorist threat, and a president who had turned out to be a relatively normal and honest man. They needed a villain, and the Montana suit had reminded them that there was one right under their noses. From a relationship-building standpoint, I'd have been better off changing my name to Manson.

"Never again," I said to myself with conviction that sounded good even through a mouthful of toothpaste. I would never again succumb to my addiction to Darius. In fact, I'd never see him again. Ever. I'd ignore his calls. And if he should come to my house, I'd jump out the window and run for the swamp that was slowly reclaiming my back lawn.

I turned the shower off and stepped out, waiting for Nicotine to shake before drying myself. I wiped the water and dog hair from the mirror and squinted at my reflection. It reminded me of college—back when I let Darius do this kind of thing to me almost every night. The greenish skin and red eyes had been a badge of honor then—a sign that one was fully exploring everything youth had to offer. Now I just looked like a drunk.

3 THE OFFICE TOWER WAS NEW: ONE OF THOSE LONG, straight glass jobs that you'd walk past a hundred times in New York but that in North Carolina looked as though it had fallen from the sky and just stuck there. I stood across the street, hiding in a minimart that had already been marked for destruction and replacement by something more in keeping with the way the neighborhood was shaping up. My guess? A law office.

I pretended to be pondering the pink snack cake in my hand, but the thought of eating it made me even dizzier than I already was. Before I could have one of those bursts of tequila-smelling sweat that would hang on me the rest of the day, I moved to the soda dispenser and began filling a cup so big that I had to use two hands to hold it.

The little store was doing a fairly brisk business. A convex mirror above my head displayed no less than ten distorted figures moving through the aisles, provisioning themselves with a kind of military efficiency: chips, Band-Aids, fizzy water, baby wipes . . .

When the long filling process was finished and I was armed with a full sixty-four ounces of Coke and ice, I made my way to the cash register, nodding solemnly at my fellow shoppers, and then escaping out into the sticky morning.

The normal haze that hung over the South this time of year had thickened into an overcast that promised a hard, hot rain and probably ex-

plained the subdued mood of the mob across the street from me. I reckoned that there were no more than twenty protesters milling around
the broad entrance of the building I worked in, and the ones holding
signs had already covered them with clear plastic. The typically energetic
and obvious chants utilizing every conceivable word that even remotely
rhymed with *murderer* had been all but silenced by the depressing
weather.

I normally avoided all this by slipping my official-sticker-covered car
into the heavily secured underground parking lot and taking the elevator
up to my office from there. This morning, though, my car was MIA—
hopefully parked safely in Darius's driveway and not wrapped around
one of Darius's trees. That left me with little choice but to mount a more
daring frontal assault.

I took a sip of Coke and started forward, conjuring an appropriately
angry and determined expression as I joined some of the more emphatic
critics of Terra, the oh-so-wholesome and eco-friendly sounding corporation formed by the recent merger of three sizeable tobacco companies,
one of which was founded by my great-great-great-grandfather. Or was it
my great-great-great-great-grandfather? I can never remember. In any
event, Terra Holding Corporation was now the largest tobacco-related
corporation in the world and was considered to be the last word in corporate evil.

As I began to break away from the group of protesters and edge toward
the elaborate glass entrance of Terra's headquarters, I heard someone right
behind me shout loud enough that I reflexively spun around and held out
my enormous Coke as a shield. It turned out to be a diminutive man of
about fifty holding a sign reading TWENTY MILLION AND COUNTING!

"Killers!"

He seemed to be yelling at the building and not me.

I smiled weakly and gave him the thumbs-up, then turned and immediately made a dash for the DMZ—an unoccupied space between the protestors and the security staff lined up in front of the building. Fortunately,
one of the guards recognized me and opened the door. I ran through to

the jeers of the protestors as they realized they'd been duped and that one of the murdering bastards had been walking amongst them.

"DON'T WE HAVE any boiling oil?"

My secretary was standing in my office with her back to me, leaning into the window to better see the protestors below. She was wearing the same slightly yellowed suit she wore every Monday and the same slightly yellowed complexion she'd worn since I'd met her. When she spun to face me, the centrifugal force smoothed her skin enough to briefly make her seem younger than her actual two hundred years.

"The windows don't open," I reminded her.

She lifted her glasses from where they hung around her neck by a silver chain and looked through them with a mix of disapproval and disappointment. The chain had been the only thing of value her mother ever owned. I can't remember how I knew that—it was the only piece of personal information I ever managed to pick up about Ms. Davenport.

"You missed your documentation meeting with Chris."

"Car trouble," I said, not expanding on the fact that the trouble was I couldn't find it.

"I rescheduled for ten."

"Fine."

She walked out of my office and, as always, I let out a long breath when she was gone.

"SO? WHAT ARE you hearing?"

I'd been getting that same question three times a day, five days a week for more years than I wanted to remember.

"'Bout what?"

Stan smirked but didn't move from his position filling my doorway. He had the general shape, size, and coloring of a large snowman and considered my cluttered office a bit claustrophobic.

"Come on, Trev. You're killin' me over here."

Because of my last name—Barnett—it had taken me a good two years to convince my co-workers that I wasn't a spy taking orders from the executive floor and it was going to take at least another two to convince them that, despite my family history, no one ever told me anything.

"What *I'm* hearing is that we're in trouble up there," Stan prompted.

"Up there" translated into "Montana" in tobacco circles. On the rare occasion when someone actually did utter the word *Montana,* it was always in a whisper.

"A guy I know knows someone who's related to a juror. Says she's always hated smokers. You'd think we could keep *them* off the jury."

I shrugged, wishing he'd leave me to my barrel of Coke and cold sweats. "People smoke, man. Always have, always will."

Stan took the extraordinary step of entering my office. He went over to a small table against the wall and snooped through the papers piled on it under the pretext of creating a space for his ample butt.

"Don't you watch TV, Trevor? It's getting worse every day. You can't turn the goddamn thing on without having to listen to people bitching and moaning about how we tricked them and how no one had any idea that smoking was addictive or dangerous. Don't they ever get tired of trying to shovel that load of crap? Did you know that Christopher Columbus wrote in his diaries that his men were getting addicted to tobacco? And that was without fifty million dollars' worth of government studies."

Actually, I was the one who'd told him that particular story. I'd always found it ironic that a guy dumb enough to think Cuba was China had all this figured out over five hundred years ago. I was pretty certain that said something profound about modern society, but I wasn't sure what.

"Didn't know that, Stan. Interesting."

I glanced over at the door and saw a crowd forming. The floor I worked on—the twelfth—was dead in the center of the building and brimming with middle managers who were about five floors too low to really know what was going on and about three floors too high to blissfully accept that ignorance.

"This is it," Stan said with some certainty. "They're finally going to get us, aren't they? They're finally going to do it. I heard that all the companies have gotten together and hired a law firm out of Germany to draw up our bankruptcy papers. People are saying that the execs are spinning real estate and other stuff off into partnerships and giving themselves ownership. They're sewing their golden parachutes, man. And they're gonna jump right before this thing crashes with all of us in it."

I looked over at the door and found four pairs of eyes staring at me, waiting for an answer. Waiting for me to tell them that none of Stan's rumors were true. That everything was going to be okay. I remained silent.

"I hear the old man's thinking about taking a job with Xerox," Stan said, still trying to bait me.

The old man he was referring to was the storied CEO of Terra. Why he was so renowned, I'd never completely figured out. Other than his habit of constantly firing people for the most imperceptible of offenses, I couldn't think of anything particularly remarkable about him. Everyone seemed certain that he'd done brilliant, important things, though, and the fact that no one could think of a specific example just added to the Great Gatsby air of mystery surrounding him. I'd only met the man in passing a few times and like every other lowly tobacco serf, I'd done a little boot-licking and then run away before he could can me because my aftershave offended him or something. Despite that, everyone on the floor seemed to think he and I were in constant consultation.

"You'd know better than me, Stan." I glanced at my watch. "Gotta run. Meeting."

I'VE ALWAYS HELD that most people could be categorized into their respective phyla and species by their reaction to trust-funders like myself. My boss, Chris Carmen, for instance, was an excellent example of a Double-Breasted Seether. In fact, he was as fine a specimen as I'd run across.

"Sorry I'm late, Trevor," he said, bustling into the conference room with a lot of superfluous motion that was calculated to make him look busier than he really was.

"No problem, Chris. It's totally my fau—"

"I was up all night." He fell into a chair and rubbed his face as though he was trying to increase the circulation to it. "Danny's got the flu and now he's given it to Karen. I couldn't make it in 'til six-forty-five, and it's put me an hour behind."

In addition to its lovely plumage, the Double-Breasted Seether could be distinguished by a high-paying job, a carefully cultivated air of martyrdom, and a complete inability to speak plainly. What Chris meant was:

"I spent the night being puked on by an infant and yelled at by my sick wife while beautiful women fed you grapes and fanned you with palm fronds. Maybe tomorrow you could have your breakfast-in-bed schedule moved up so you can get to work on time."

And while I have to concede that a near-terminal hangover is a poor excuse for missing a meeting, I figured his life wasn't all that bad. I'd met his wife—she is beautiful, intelligent, and clearly loves him. I assume his kid was cut from the same cloth.

"Now, Trevor, I realize we had a problem with the servers and that cross-referencing ground to a halt last week. But still, we're three weeks behind on the ten-forty-eights and two weeks behind on the ten-fifty-threes. With the servers back up, we've really got to push to jump this thing back on schedule. What we don't need is this snowballing and putting us behind on the sixty-fives . . ."

I'll spare you rest of this conversation and cut to the really sad part: My job isn't even as interesting as it sounds.

I'd been put in charge of computerizing the hundreds of tons of written documents generated by the tobacco industry over its long history. This meant scanning them into a digital format, checking for errors, cross-referencing, hypertexting, classifying, massaging, and so on. Basically, distilling two hundred years of politics, lawsuits, monopolies, and death in such a way that information could be efficiently and exhaustively queried by the handful of people who had the authority to do so.

The joke went that the industry had told so many lies that it was impossible to keep them straight without the aid of lightning-fast Pentiums. Obviously, it's one of those jokes that's less funny than it is true.

How had I landed this exciting career? The long answer is kind of complicated, but the short answer is simple: A person with my job description was in an excellent position to stumble across a document so incriminating and horrifying that it would destroy the tobacco industry and make the world grind out their butts in a unified and heartfelt show of disgust. Of course, the whole idea of that is laughable, but our executives hadn't quite wrapped their minds around the fact that there was absolutely nothing even remotely conceivable that could make us look worse than we already did.

In light of management's somewhat fanciful notions about our public image, I became an obvious choice for the job. The theory was that it would be inconceivable for someone with my family history ever to become a whistle-blower.

"So what do you think, Trevor?" Carmen said, tapping his fingers nervously on the file in front of him. "I'd jump in and give you a hand but . . ." He let his voice trail off.

And that was another thing that Carmen hated about me. I had the company's highest security rating—probably a good five levels above his. He simply didn't have access to the documents he'd need to get to in order to help me.

"Don't worry, Chris, I'll—"

"Trevor?"

We both twitched at the sound of Ms. Davenport's voice. Chris was afraid of her, too.

I twisted around in my chair and saw her leaning in the partially open door. "Yes, ma'am?"

"Have you finished your analysis of the new surgeon general's report?"

My eyes narrowed in an expression of pain that I really did feel. It was sitting half finished on my desk.

"The board meeting's tomorrow at ten," she warned. "Science already has theirs done and I hear Legal's going to finish this afternoon. If you're going to beat them, you don't have much time . . ."

The deal was that the department that finished its report last was responsible for delivery. A terrific incentive to get your work done in a timely manner given Paul Trainer's habit of shooting the messenger.

"I'm on it," I lied. "Thanks, Ms. Davenport."

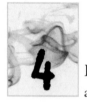**4** DESPITE FILLING MYSELF WITH HALF THE McDon-
ald's menu, a handful of B-complex tablets, and an inhuman
120 ounces of Coke, my hangover just wouldn't give up. The only tangible effect of my diet regimen was enough trips to the bathroom that the people on my floor were becoming concerned about the condition of my prostate.

On the surface, I should have ignored my suffering, buckled down, and finished my pointless analysis of the surgeon general's report in plenty of time for the board meeting tomorrow. But I'd already lost my race with those prima donnas from Legal, and I was starting to fade badly.

The task I'd been charged with was differentiating this SG report from the prior ones and finding a place for it in the context of the new millennium. Theoretically, that shouldn't have been too tough. Other than the more refined scientific methods, larger statistical samples, and graphs that were now in full-color 3D, the report was just another reiteration of the same demands: additional education (as though no one knows smoking is bad for them), federal and private insurance to cover nicotine patches (which don't work), and starry-eyed testimonials about the much more draconian antitobacco policies of the Canadians (who smoke the same amount we do) and the Europeans (who smoke, well, like chimneys).

In my mind, though, it was the last paragraph that really showcased the SG's unique genius. I quote: "The impact of these various efforts, as measured with a variety of techniques, is likely to be underestimated because of the synergistic effect of these modalities." Now, I don't know what a synergistic effect of a modality is, but I'd swear that this says: "No matter how many Ph.D.s tell you people still smoke like fiends, and no matter how many people you see lighting up, and no matter how many billions the tobacco industry grosses—trust us: We're winning the war against smoking."

Now about three-quarters finished, my analysis had turned into an exercise in transforming what was really a single sentence of information into twenty-five hundred words. I'd had to use every trick in the book to do it: even sinking to word processor hijinx and replacing the word *too* with *as well* throughout the entire document. You know you're desperate when you start playing those kinds of games.

Eventually, my growing case of writer's block had become bad enough to demand drastic action. And in this case, that action was accepting my co-workers' invitation to happy hour.

The restaurant had sprouted up only a few months after the completion of Terra's headquarters and was more or less what you'd expect: lots of vaguely clever old signs advertising (nontobacco) products from the fifties, waitstaff in striped rugby shirts, a free buffet made up of the stuff that wasn't quite good enough to serve to paying customers.

Since the Montana suit had started, this place had become more and more popular—filling quickly with nervous people who needed some after-work sedation and a dose of the latest gossip. As near as I could tell, everyone had a cigarette in their hand whether they smoked or not. A sign of solidarity.

"You had to run right through them?" Stan said, lifting a limp roast beef sandwich to his mouth. "Damn! I heard a woman from accounting got a dead bird thrown at her. Bet they poisoned it so that they could use it for ammunition—those people have no sense of decency, man. You should have just gone and gotten your car—you're lucky you made it through them without getting recognized. They've got files on all of us, you know. They know everything about you, me, and everybody else."

Despite all evidence to the contrary, Stan had bestowed on the anti-tobacco lobby Godlike omniscience and a sort of cold war–era Soviet evil.

"It's all on the 'Net," he continued, cheeks so full of beef and his face so red that he looked like a trumpet player trying to hit a high note.

"What is?"

"The files, man! You just have to know where they are and have the passwords. Then it's all there: Where we live, how much we make, our high-school transcripts . . ."

"Have you ever seen them?"

He shook his head and dabbed at his mouth with a greasy napkin. "They keep it too well hidden."

Stan was the head of a small marketing group known within the company as the Ministry of Misdirection. The focus of the Ministry was simple but surprisingly elegant. Because it was getting harder and harder to make 300,000 smoking-related deaths just fade away every year, it had been decided to try to divert everyone's attention instead.

It was Stan's mission to promote awareness of the other general dangers of being alive: handguns, alcohol, lead paint, your cell phone giving you a brain tumor—that kind of thing. He spent his days egging on the country's more militant consumer advocates, providing them with access to Terra's considerable marketing expertise (through even more innocuous-sounding subsidiaries, of course), and using his contacts to get the media to splash across the nation's TVs the carnage caused by such things as slippery bathtubs and airbags.

Whenever there was a school shooting and everyone else was asking, "How many kids were hurt?" Stan always wanted to know if the shooters were drunk. Guns, booze, and the death of children neatly wrapped up in one dramatic incident were the Holy Grail for a man in his position.

The crowd filling the bar shifted in a way that would allow me a more or less straight dash to the bathroom, and I used that strategic moment to take a single, courageous pull on my beer. My stomach rolled a bit, but all in all, it went down pretty well.

"You still dating that girl?" Stan asked, poking at a distracting piece of beef fat hanging from between his teeth.

"What girl?" I took another swig from my beer and this time suffered no gastrointestinal reaction at all.

"Blond skinny thing. Had one of those highbrow names . . ."

"Morgan."

He flicked his lips with his tongue, undoubtedly imagining her long, naked body framed by a set of satin sheets.

"I don't want to be a jerk, Trevor, but I gotta know. What's she like? You know . . ."

While not a Seether, Stan was another person with a sweet wife and cute kids who had convinced himself that I lived the life of a rock star. I didn't answer, instead taking a third pull from the bottle in my hand. This time the icy liquid flowing down my throat actually made me feel better.

"Come on, Trevor. I swear, I'll never ask you anything like this again. But you're killing me here."

The truth was that sex with Morgan had been as sterile an activity as I'd ever participated in. But Stan didn't ask for much, and it just didn't seem right to disappoint him.

"Crazy, man. Crazy."

"I knew it. It's always the stiff, bi—" He caught himself before he said "bitchy ones."

"Anyway, must be all that repressed energy . . . So? You still going with her? Is it serious?"

Honestly, I'm not sure we'd ever been "going together." She'd wanted to rebel and, as the black sheep of a "good" southern family, I was just lunatic fringe enough to make her daddy angry but not so lunatic fringe as to make him furious. When it had become obvious that he didn't care one way or another, she'd gotten bored with me pretty fast.

"Couldn't keep up," I said. "Seriously. She had me up 'til three A.M.

seven days a week. It was getting so I couldn't drag myself out of bed in the morning."

His fat cheeks ballooned again as a wide grin spread across his face. I lit a cigarette and took a delicate puff.

"So what're you hearing about Montana?" he asked me for a record seventh time that day. "What's the inside track?"

I shrugged and he leaned into me, lowering his voice so as to make sure none of the people around us could overhear. "Come on, man! We're friends, right? We've been friends for what, five years? I've got kids to feed and a wife who doesn't like to work. Should I be looking for another job?"

"I honestly don't know, Stan. If I did, I'd tell you. I swear."

"It's bad isn't it?"

I drained the rest of my beer and waved at the bartender to bring me another. "If you've got an offer from another company on the table, Stan, you might want to think about taking it."

I DIDN'T MAKE it back to my desk until after ten.

The quick medicinal beer had turned into five medicinal beers and about thirty not-so-medicinal Buffalo wings. I fell into my chair, rubbing my bloated stomach and staring at my reflection in the windows of my office. I looked better—the green hue of my skin had turned to a kind of a high-blood-pressure pink and the gleam of hangover-induced sweat was gone from my skin, leaving my image a little hazy and lifeless. A more normal state for me.

My partially finished analysis of the new SG report was up on the computer screen and I turned toward it, having a hard time bringing the words into focus at first. When my eyes finally adjusted, I found that even with a healthy buzz, it still had the feel of a poorly written high-school book report.

"What do you want from me?" I said to the empty office. No answer.

The amazing thing was that this job had seemed great when I took it. In fact, "great" didn't even capture my initial enthusiasm. "Perfect" would be more accurate. I mean, it involved history—something I was genuinely interested in—it afforded me near total autonomy; there was really nowhere to get promoted to and therefore the position boasted an almost complete lack of accountability and politics; and finally, I wasn't physically involved in the production or marketing of cigarettes. All I did was neatly arrange papers that described things that had already happened.

What I hadn't counted on was the fact that spending your life doing something tedious, pointless, and endless had certain drawbacks. And the truth was, no matter how much I denied it, I was still a cog in the machine. Worse, I was a cog that sat around rationalizing away my involvement with meaningless technicalities. At best a not-so-original industry pastime, at worst a century-long industry tradition.

I leaned my forehead against the edge of my desk and mumbled a few obscenities.

Nine years—my entire adult life after college—that's how long I'd worked there. Actually, that isn't entirely true: I'd quit twice. I can't remember the catalysts for those uncharacteristically bold acts anymore but I definitely can remember marching into Chris's office, slapping my resignation on his desk, and walking out. I also remember crawling back two days later and asking for my job back. He hadn't bothered to look surprised but, to his credit, he hadn't made me beg.

Perhaps a little explanation is in order here:

The trust my grandfather created for me consisted of nearly 200,000 shares of tobacco industry stock and provided me with two fundamental benefits. The first was that upon graduation from college, I would be entitled to annual payments based on an arcane formula involving capital gains, dividends, and inflation. The second was that at age sixty the trust would be fully distributed to me. There were a number of catches of course, but the main ones related to my employment. In order to receive the annual payments I had to have worked for a tobacco company for the entire preceding year and in order to get the final distribution I had to have spent my entire career working for the industry.

With this and my father's similar but much more lucrative trust, my grandfather had guaranteed a motivated dynasty of Barnetts toiling in the tobacco fields for two more generations. In fact, based on the average age of death when the documents had originally been drafted, he'd intended for us to die at our desks, having not lived long enough to ever get our hands on his money.

What no one who wasn't in my position could comprehend was the incredible power of that elusive payout. I'd known it was there from before I was old enough to understand what it was. Over the years, it had taken on an almost human quality—like a kindly relative. "Don't worry, I'll take care of you," it whispered in my ear. "You'll never really be on your own."

As time went on, though, I came to realize that it was a trap. Actually, *trap* might be too strong a word since it was so easily avoided. But trapped I was.

This was all the more pathetic when you consider what my trust was doing for me. I hadn't seen a dime of it since the industry had started to tank two years ago. My last stock statement had shown a total value of just over a million dollars. And after the Montanans got through with us, that would likely plummet even further. Taking inflation into account, I'd probably have just enough to go out to dinner on my sixtieth birthday. But only if I went alone—which was more than likely the way things were going.

"Fuck you, too!" I shouted at the empty building.

The acoustically engineered cube farm outside my door absorbed the sound with the ruthless efficiency I'd come to expect of all things tobacco. I jumped to my feet, swaying dangerously from the alcohol in my system, and shouted louder. "You heard me! Fuck you!"

What the hell was I doing with my life? Darius was making hundreds of millions of dollars a year. Einstein had already developed the theory of special relativity at my age. Half the movie and rock stars in the world were younger than me. There were thirty-two-year-old doctors and thirty-two-year-old lawyers and thirty-two-year-old congressmen all out there looking down on me. And why not? I'd spent the last nine years doing absolutely nothing, in anticipation of being rewarded for that laziness when my life was mostly over. Suddenly, I started to remember what had prompted me to quit the first two times.

I threw myself into my chair and closed my moronic summary of the surgeon general's report in favor of a blank page that would contain my latest resignation letter. I began typing quickly, promising myself that this would be the last time. Weak economy or no weak economy, I'd hit the streets and find myself a real job. I might end up cleaning out garbage cans for five bucks an hour, but damn it, I'd be doing it on my own.

I'd made it about halfway through the rather tersely worded letter when I found myself slowing down. What if finding a decent job was hard? What if it was *really* hard? Terra would still be here and so would Chris, sitting there in his office waiting for me to come crawling back.

And what if I did? What would my future be? How long could Terra hold out against a court system bent on its destruction? What if it was a long time? What if instead of going out with a bang, the company just slowly crumbled? What if I came back and the company folded when I was forty? Or fifty? It would be too late for me then. Too late to do anything with my life.

I started typing again, hard enough that the clack of the keys bounced around my office a bit before being sucked into the void beyond. Not this time, I promised myself. This time, I wouldn't come back . . .

Would I?

I slowed again and then stopped, finally leaning back in my chair and staring at the computer screen. How long I sat there, I'm not sure.

For some reason, I sensed that this was it—the fork in my road. Either I decided to go it alone now or I just resigned myself to fading away along with the industry my family had started.

I thought about the board meeting tomorrow and I thought about Paul Trainer, Terra's egomaniacal hatchet man of a CEO. Then I deleted my rather verbose letter and replaced it with a single sentence, centered on the page.

After admiring the concise eloquence of my new SG report analysis, I printed it and tucked it lovingly it into the folder from Legal.

In less than twenty-four hours, I'd be free.

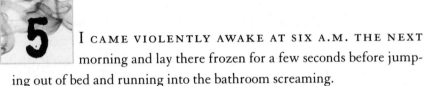**5** I CAME VIOLENTLY AWAKE AT SIX A.M. THE NEXT morning and lay there frozen for a few seconds before jumping out of bed and running into the bathroom screaming.

"Shiiiiiiiiiiiiit!"

I nearly ripped the shower door off the wall, turned on the water, and threw myself beneath the icy stream. I staggered back until I hit the cold tile behind me and stood there taking shallow, shaking breaths.

"Okay," I said aloud to myself. "Relax. Just relax and think for a minute."

Nicotine wandered in and flopped down next to the toilet, baring her teeth in a wide yawn. She didn't seem sure what to make of the situation but clearly didn't want to miss anything interesting.

I put my hands in front of me to block the still-frigid water and tried to calmly assess my situation.

Physical: Not too bad. My happy hour antics and a lack of sleep had combined to extend my hangover, but it was one of those friendly ones where your skull feels big and empty but nothing really hurts. My heart was pounding out of control, but that fit better under the next category . . .

Psychological: My emotional state would best be described as desperate, uncontrolled panic combined with the sensation of being completely alone in the universe.

Employment: Soon to be nonexistent. When Paul Trainer got my one-sentence summary he was going to completely melt down. He'd once fired a twenty-year veteran of the company for parking partway in his space. I'd be lucky if he didn't tell Security to work me over before they threw me out in the street.

Financial: Broke. My savings account continued to dwindle as I raided it every month to pay the mortgage that was supposed to be covered by my trust distributions.

Prospects: None. When Trainer got through blackballing me, I'd have to move to another planet to get a decent job.

"Shut up," I ordered myself, though I hadn't been talking aloud. The water had finally warmed up a bit and I took a hesitant step forward, letting it pound on the top of my head. "It's okay, it's okay . . . This is what you wanted . . . You can do this."

It seemed like a stretch, but I forced myself to modify my assessment.

Prospects: Limitless.

I WAS TRAPPED at the back when the elevator doors opened, and I almost knocked an old lady over when I forced myself forward.

Despite the fact that it was only ten after eight, I crossed the floor at a full run—dodging around knots of early-morning gossipers and ignoring their obvious curiosity. When I rounded the corner, I saw Ms. Davenport's face hovering just above the wall of her cubicle, scanning the horizon like a sailor worried about icebergs.

"Why are you late?" she said as I skidded to a stop on the low-pile carpet. There was a strange whine in her voice that I'd never heard before and that softened the normal authoritarian finality of it.

"I was in here all night working on my report," I lied. "Besides, I'm not late. It's only quarter after eight."

That seemed to confuse her, so I glanced into my office and checked my statement against the Marlboro clock hanging above my desk. It read eight-fifteen exactly. Plenty of time to tear up all evidence of last night's bout of temporary, alcohol-induced insanity and deliver to the board the twenty pages of repetitive drivel still saved on my hard drive. It wasn't a piece of writing that would win me any accolades, but it wasn't bad enough to get me fired either.

I feinted right, then shot around her to the left, going for the coffeepot on top of a bank of file cabinets. "You'll have my report in five minutes, Ms. Davenport. Sorry it's last minute. If you need help with the copying, I'd be happy to pitch in."

"What report?"

"The *board* report," I said, trying to sound casual. "You know, the one you've been on me about for the last two weeks?"

That seemed to perplex her even more, and I wondered if she'd had one of those mild strokes people her age got. According to the new surgeon general's report, smokers were thirty-one percent more susceptible . . .

"Are you feeling all right, Ms. Davenport?"

"I didn't see a copy of it!"

I shrugged and took a sip of coffee. It was too hot and I sucked in a breath, trying to cool it. "It's on my computer. I know I should have printed off a copy and left it on your desk last night, but I was beat."

I started to turn toward my office, but her expression stopped me. It wasn't the expression itself, actually—more the fact that it was completely frozen to her face.

"Seriously, Ms. Davenport. Are you oka—"

"How was I supposed to know it was on your computer!" she said at nearly a shout. "There was no way for me to know!"

I took a step back, increasing the distance between us in case she decided to attack. "Um, no problem, ma'am. Tell you what, I'll just deal with the copying myself and run them upstairs when I'm finished—then you don't even have to deal with it. Ten o'clock, right?"

She began playing with the chain holding her glasses around her neck. "The . . . the meeting got moved up."

I blinked a few times. "What?"

"It started half an hour ago. I tried to call you, but you weren't answering your phone."

As I let that process, I felt the mild queasiness caused by my hangover surge dangerously. "Are you telling me that . . . that the entire board is upstairs meeting about a set of reports they don't have?"

Ms. Davenport shook her head. "The reports from Legal and Science were on your desk. I had the mailroom run copies while I tried to find yours. I looked everywhere, Trevor. I even tried your computer, but I don't have access to a lot of your files . . ."

I was barely able to keep myself from hyperventilating as my mind filled with images of the directors of one of the largest corporations in America—men and women who were captains of industry, former high-ranking politicians, and worse—sitting around discussing everyone's report but mine.

"Trevor, I'm sorry. I didn't know what to do . . ."

I made a mad dash for my office and started jabbing desperately at my keyboard, pulling up my original report and printing it as I mumbled pathetically to myself. "It's only been a half an hour—they're probably not even through the first two reports yet, right? I'll . . . I'll just tell them it was a computer glitch. They won't say anything. Half of them have probably never even turned on a computer . . . It'll be okay . . ."

I ONLY WAITED about ten seconds for the elevator before running for the stairs with twelve hastily stapled copies of my report under my arm. Normally, the elaborately constructed and richly decorated stairwell was my favorite part of the building. The walls were covered with beautifully framed, chronologically organized copies of important paintings. The theory had been that if you made the stairs the nicest part of the building, people would use them and the exercise would help raise Terra's workers' life expectancy above that of sub-Saharan Africa.

I bounded through Impressionism on pure adrenaline and managed to keep my momentum through Cubism, but by the time I got to the abstract multimedia stuff I was wheezing like an old man with asthma. Fortunately, no one but me ever came in here and so there would be no stories of me doubled over on the steps, coughing and gagging in front of the entrance to the executive floor. When I got my breathing more or less under control, I punched a few numbers into a keypad next to the door and staggered through.

No one seemed to notice as I jogged through the elegantly paneled halls toward the boardroom. It was like sweaty, panicked thirty-somethings ran through there every day.

"Can I help you, Trevor?"

Susan Page, Paul Trainer's assistant, smiled benevolently.

"I need to . . . um. I need to talk to the board, Susan. I've got to—"

"Go on in. They're expecting you."

My breath caught in my chest for a moment. "What do you mean?"

"Uh . . . I mean, they're expecting you."

I nodded stiffly and walked down the hall, stopping in front of the double doors that led into the boardroom but not entering.

This wasn't a cop-out, I told myself for what must have been the hundredth time. The smart move was to try to save my job at Terra and then start working on my resume tonight. I mean, getting fired and pissing off Paul Trainer wasn't exactly something that was going to impress a prospective employer. No point in burning your bridges behind you. Right?

"Trevor?" Susan said, peeking around the corner. "You can go right in."

I smiled weakly and pulled the reports from under my arm, relieved to see they weren't damp with sweat.

For some reason, I was afraid to open the door too wide and I'm sure I ended up looking like an idiot as I squeezed my considerable bulk through too small a crack. In the end, though, I made it and closed the door quietly behind me before backing myself into a corner.

No one seemed to notice my arrival, concentrating instead on an argument going on between Paul Trainer and a man I didn't know. I was a little too overwhelmed to track on it, instead using the time to go over a little computer mumbo jumbo that might explain the tardiness of my report.

"I don't give a shit what the goddamn media is saying!" Trainer shouted. "And I'm not listening to any crap about journalistic integrity because we all know there's no such thing and never has been. This is about the same thing it's always about. Money. Plain and simple."

"Come on, Paul. The tide's turning further and further against us, and the media's along for the ride. I agree with you that we can use our advertising budgets to pressure them, but we've got to be subtle. We can't just cut off organizations that come out with negative stories about us. It would backfire. I guarantee it."

Trainer waved his hand in the way powerful men do when they know they're wrong. Then he looked directly at me.

"Trevor Barnett," he said in that old-timey drawl of his. "I'm so happy you could spare us some of your time."

"Yes, sir. I'm afraid there was a fatal hard-drive crash and my backup on the network—"

"What the hell are you talking about, son? I swear to God, I can't understand a thing you kids say anymore."

For some reason, the polite titter in the room made it occur to me that I should step out of the corner I'd folded myself into. "What I meant to say was—"

Trainer cut me off. "Because as near as I can tell, you seem to have no problem speaking plain, concise English."

I was confused for a moment and then felt the heat in my face when I realized I was being made fun of. My nonexistent report was the height of conciseness. Ha, ha, ha. Good one. My eyes shifted to either side of the table that Trainer was sitting at the head of, scanning the serious, aging faces poking out of expensive suits.

A few years back, a columnist from an uncharacteristically critical Christian magazine had compared Paul Trainer to Satan and bestowed the elaborate names of lesser demons on the rest of the board. At this moment it wasn't hard to imagine them that way.

Trainer flipped open one of the folders in front of him—the Legal analysis judging from the faux leather cover—and pulled a single sheet from it.

"Smoking is still real bad for you," he read. "And we have no idea what we're going to do about it."

I'm not generally what you'd describe as a dense person, but in my current rather stressful situation, it took a few moments for me to recognize my own words and to connect everything in my brain. When I did, I felt my face turn a shade of red that would have shamed a tomato. Those jerks in the mailroom had copied the page I'd stuffed in the Legal folder and put it in the board's package.

I looked at my shoes and then glanced up into the face of the company's general counsel—Beelzebub to the *Northern Christian Weekly*. Beneath thinning, mousy brown hair, his face was redder than mine. Maybe it was just the lighting, but I'd swear his anorexic little lips were turning blue.

"Are you all right, Dad?" I said before I could catch myself.

He jerked in his chair as though he'd been bit, shot his colleagues a look that suggested he was surprised to discover I was his son, but he never acknowledged me. I stepped back into my corner again.

Trainer regained control of the meeting by grabbing the Science report and flipping through it with loud flicks of his yellowed nails. I just stood there feeling weaker and weaker. This was all a stupid mistake. Surely, I could explain—I could say that the page tucked into the Legal report was there by mistake and had actually just been a statement of theme for the excellent report I now held in my sweaty palms . . .

"Trevor," Trainer said finally. "I want to congratulate you."

His voice held no sarcasm but obviously the statement itself did. The board members surrounding the table had the good grace not to snicker—probably out of respect for my father's plight. They undoubtedly all had children who were disappointments, too.

"In documents so thick they could not be penetrated by a small-caliber bullet," Trainer continued, holding the Legal and Science reports up in front of him, "you have managed to write the only thing that makes a bit of goddamn sense to me."

There was silence as we all waited for the punch line.

"I was starting to think I was going crazy, son. I'd like to thank you for proving me sane."

More silence. I looked over at my father for some clue as to what was going on, but he didn't seem to understand either.

"Yes, sir," I said when I realized Trainer expected a response. "Thank you, sir. Now I'll let you get back to your meeting. I'm sorry to have bothered you."

I'm embarrassed to say that I slid along the wall until I could reach the door and squeeze through it.

Trainer's eyes were on me the whole time.

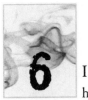**6** I SAT IN MY OFFICE, ELBOWS ON DESK, HEAD IN hands, for so long that my neck started to cramp. I tried to will myself to sit up straight and start cleaning out my desk—to show a little dignity—but all I could manage was to shift to a slightly more comfortable position.

I'd replayed Paul Trainer's words a hundred times and each time they'd taken on a little more venom. Why hadn't he fired me on the spot, as was his custom? Simple. He was too mad for that. He perceived that I—a useless thirty-two-year-old trust-funder who'd never done anything worthwhile in his whole life—was disrespecting him. There was no way a simple firing was going to satisfy him. He wouldn't rest until he completely destroyed my life.

After another ten minutes of conjuring up elaborate mental images of the things Trainer was planning for me—many of which hadn't been legal in two centuries—I finally stood and took a few deep breaths.

"Pull yourself together," I said to the empty office. "What's done is done. It's time to think about the future. What are you going to do?"

I walked over to open my door but thought better of it and went to stand in front of the window instead. The protestors were still milling around below like colorful, sign-carrying ants. I imagined I could hear their shouts, though the building's exterior soundproofing had been designed with them in mind.

I assumed that the world wasn't going to give me a lot of sympathy. Why should it? I was an able-bodied, reasonably young American with a college degree and a few dollars in the bank. The world was mine for the taking, right?

Then why was I already trying to figure how I could grovel convincingly enough to get my father to go to Paul Trainer and intervene on my behalf? I mean, I didn't need to keep this particular job, right? They could put me somewhere Trainer would never see or hear from me again. I could work in the . . .

Christ.

At that moment, could I have been any more pathetic? My father had never run to my rescue in his life. By now, he was probably halfway finished tossing my baby pictures into that stately fireplace of his.

There was a knock at my door, and I spun around. They were here. Security was here to escort me off the premises. I'd seen it done before—a guard on either side, not touching you but close enough that they could grab you if you started to make a scene.

The knock came again and I stood, squaring my shoulders and thrusting out my chin. It was true that I didn't have much to hold my head high about, but that didn't mean I couldn't. It didn't mean I couldn't walk out of there clinging hopelessly to that damp spark of pride I'd been unsuccessfully trying to fan since I was a kid.

"Come in."

The knob started to turn . . .

"What happened with the board meeting?"

It was just Stan.

"Jesus, man," I said, reaching for the back of my chair for support as I felt myself sag at the knees.

"What?"

"Nothing. Okay? Nothing?"

"What's wrong with you, Trev? Did the board beat on you?"

I shot him a hostile look, paranoid that he knew exactly what had happened and that he was just playing with me.

"You're getting to be a secretive bastard, you know that?" he said, pushing the door completely open. "I heard from one of the secretaries up there that you were in the meeting for a pretty long time. You're working on something for them, aren't you? And don't try to tell me they just wanted to talk to you about the goddamn surgeon general. I may have been born at night, but not last night."

I tried to will him out of my office, but he just settled a pudgy shoulder against the doorjamb.

"I just went up there to pass out my report, Stan. That's it."

He smirked. "I heard that the execs looked like Trainer had taken his nine iron to 'em when they got out of there. And I heard he spent special time with your old man."

I winced, picturing my dad having to sit through half an hour of his boss telling him what a loser I was and speculating on his failings as a father. I stared at the floor and didn't respond, hoping that Stan would eventually take the hint and close me back up in my office so I could wallow alone. He didn't, though. I could still see his shoes.

"Okay, enough of this," he said finally. "Is it true?"

"Is what true, Stan?"

"That your whole report was one sentence? 'Smoking is damn bad for you'?"

Not only was the story out; it was already being exaggerated. I saw no point in denying it, though.

"I guess. More or less."

"Shit, man. You really do lead the good life. I would kill to have seen their faces. What did they do? What did they do when they read it?"

"Nothing." I fell into my chair. "Nothing at all."

He started shaking his head appreciatively but then froze halfway into the motion. "Thanks for the input, Trevor," he said abruptly. "Let me work on that and I'll get back to you with the results."

"What?" I said, but he was already out of earshot—hustling back toward his cubicle with the unnatural gait of the overweight speed walkers I sometimes saw burning up the sidewalk in my neighborhood.

I assumed that he'd seen Terra's storm troopers bearing down on my office, so I stood and squared my shoulders again. I was working on my chin thrust when Paul Trainer came strolling through my door.

"Never been down here," he said, brushing by me. His old bones protested audibly beneath a thin layer of skin as he sat down in my chair. "Cubicles! What an insidious invention. Makes me feel like a rat in a maze. Really. I get nervous that I'll get trapped and starve or something."

I stepped back slowly, not listening to what he was saying as much as wondering what he was doing in my office. It finally occurred to me that he was so enraged that whatever he was going to do to me, he wanted to do it personally.

He crossed his chopstick legs and smoothed an imaginary wrinkle from expensive blue slacks. "I'd like to ask you a favor, Trevor."

What the hell was I doing standing there like an idiot when I could be prostrating myself before him?

"Mr. Trainer, I want to apologize—"

He made a cutting motion across his throat, silencing me.

"Listen, I need you to go meet with our attorneys in Montana. I need you to do it tonight. The trial's pretty well under way, and I've got to tell you that I don't understand what's going on up there. Whenever I ask a simple goddamn question, I end up with fifty pages of ass-covering in Latin. Lawyers are like politicians: The truth is whatever they can convince themselves of when they get up in the morning. You don't seem to have that failing."

Was this a joke? No, not a joke. He was setting me up somehow.

"You can talk now," he said.

"Um. I, uh . . ."

"Actual words would work better for me, Trevor."

"I'm not a lawyer," was all I managed to get out.

"Neither am I. Got a goddamn English degree. Did you know that?"

"No, sir."

"Based on what you wrote about the surgeon general's report, it looks to me like you know how to make a simple point. That's what I need: I need you to tell me in plain English where we stand up there."

"I've got a meeting tonight that I can't miss," I said, my mouth working completely independently of my brain.

"Have I given you any reason to believe I care about your other plans?"

"Getting a flight's going to be—"

"You say 'can't' a lot, don't you, Trevor? Or maybe it's just me. Do you have some problem with me? Did I cross you in a prior life? Is there some reason you won't do me this simple favor?"

"No. It's just . . ."

"Good. Talk to Susan. Take my jet."

He stood and started for the door, but then paused directly in front of me. Trainer was probably a good eight inches shorter than me, and I tried to look down at him in a way that wouldn't make it seem like I was suggesting he was short.

"Maybe something slightly more verbose than your last report," he said, tapping my chest with a crooked finger. "No. Better yet, why don't you give me an oral presentation."

He ducked around me but stopped again, this time in the doorway. "Show me you've got some common sense, Trevor. Okay?"

Then he was gone, slamming my door shut behind him and leaving me to try to figure out what had just happened.

Maybe he'd really liked my analysis of the surgeon general's report.

No. That was stupid. There was no way. This was just part of his plan to torture me before going in for the kill. I remembered what he said about lawyers being like politicians. Was he angry with my father for his failure to keep the courts from finding a way to tear the industry apart? Was that why he was sending me to Montana? Was it a calculated insult to my dad? What better way for Trainer to show distain for his legal team than to send T. Edwin Barnett's screw-up son to check up on them?

I had no idea how to take things at this level. The subtle back-and-forth of office politics was way beyond me. What I didn't need to do, though, was make my relationship with my father any worse than it was. But what choice did I have? Trainer had been clear that he wasn't really asking me to go. He was telling me to go.

I sat down again, working to clear my mind. There was no point in my trying to outmaneuver anyone—it wasn't my thing. I had no choice but to just go along.

That, more than anything, *was* my thing.

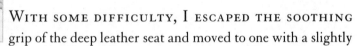

7 WITH SOME DIFFICULTY, I ESCAPED THE SOOTHING grip of the deep leather seat and moved to one with a slightly different curve to it. Then to another, and another, until I'd tried them all. Every one seemed more comfortable than the last. It didn't matter, though, since I was having a hard time sitting. I went over to a polished wood bar and tried to figure out how to free a bottle of cleverly secured Bushmills, but it turned out to be a little too cleverly secured and I quickly lost interest.

The floor dropped beneath me a bit and I felt an unfamiliar tickle in my stomach as I reached out to steady myself. I wanted to say something, to ask if that kind of thing was normal, but didn't want to come off as a complete bumpkin so I kept my mouth shut.

I was the only occupant of the surprisingly spacious jet—except for the pilot and copilot, of course. Just me, the thick carpet, the expensive liquor, the plush leather, and an impressive collection of silver ashtrays.

My family, for reasons that now seemed obvious to me, had never traveled much. In my thirty-two years, this was only my second time on an airplane. But now here I was, taking an unfamiliar mode of transportation into completely uncharted territory for unfathomable reasons.

The key to staving off panic was keeping my mind occupied, I decided, so I walked up to the cockpit door and peered at the men inside. They didn't seem to be doing much. I stood there and tried to figure out what all the dials, buttons, and switches were for until the pilot turned and graciously asked me if there was anything he could do for me. I shook my head and went back to retest the seats.

Another fifteen minutes and I'd done everything there was to do but fly the plane. I tried to fill my mind with a magazine—*Corporate Jet Review* or something equally esoteric—but it wasn't long before the fact that I was on my way to interrogate a bunch of $500-an-hour lawyers about a subject I knew almost nothing about started to creep back into my mind. What was I going to ask them? What did Paul Trainer want from me?

I pulled my cell phone from my pocket and after clearing it with the pilot, dialed a number from memory.

"John O'Byrne."

"Hey, John, this is Trevor."

"Trevor? I can barely hear you. You're not on a cell are you?"

"Yeah, I—"

"Jesus! Are you crazy? Do you know how easy it is to monitor one of those things?"

Honestly, I'd never given it much thought.

"I just wanted to tell you that I can't make it tonight, John."

"But we've got the meet all set up. It has been for weeks."

Why was it that *meetings* were always aboveboard but *meets* had a sinister edge to them? John loved all this cloak-and-dagger stuff, so I tended to play along to whatever degree possible.

"Unavoidable," I said. "Sorry."

"It's okay. No problem. I'll take care of the reschedule and contact you the normal way. And Trevor? For God's sake, use a land line next time!"

I cut off the phone and was in the process of stuffing it back in my pocket when it started to ring. I didn't immediately answer, instead I just stared down at it for a few moments. Hardly anyone ever called me and

when they did, it was never on my cell. Almost no one had the number . . .
I smiled. Darius did.

I was trying to figure out a way to work in that I was on a corporate jet
as I picked up. Maybe something like "What? You're breaking up. Maybe
it's turbulence from the corporate jet's engines." No, too obvious.

"Hello?"

"What the *hell* was that all about?"

An uncomfortable jolt of adrenaline forced its way through my body.
"Dad . . . Uh, how are—"

"I've got personal-injury suits going in every state, including a class action in Montana that's the most dangerous thing the industry's ever faced
and you're playing schoolboy pranks in the boardroom? Do you know
how that makes me look?"

"I—"

"Do you?"

It was the longest conversation we'd had in over a year, breaking the
record set by a meaningless back-and-forth about the condition of his
lawn and the prospect of rain.

"I'm really sorry, sir. Honestly, it was just a stupid mistake. A series of
them, actually. It started when—"

"I don't want to hear excuses, Trevor. That's all I've gotten out of you
your whole life. Jesus Christ! You've been handed everything! All you
have to do is wake up in the morning and not fuck it up! But you can't
even manage that, can you?"

"No, sir," I mumbled. "I'm sorry. I—"

"Sorry's not good enough. You get your ass up to my office right now
and we're going to have a talk about how you're going to behave at this
company going forward. You're going to start toeing the goddamn line
here, Trevor, or you're going to find yourself on the street. Do you understand me?"

Suddenly I didn't want him to know what I was doing. It seemed more
and more likely that Trainer was purposefully insulting my father through

me. I took a quiet breath, trying to center myself a bit. I had an unfortunate tendency to do this half-stutter thing when I was lying and the cell static wasn't bad enough to cover it up.

"I'm not in the office right now, sir. Can we do it tomorrow? Maybe in the afternoon?" Not really a lie, and I'd actually spit it out pretty smoothly.

"Already gone for the day, Trevor? It's goddamn three-thirty!"

"Dad—"

"You seem to be putting a lot of stock in your name. It's not going to get you as far as you think. I guarantee that Trainer's going to come after you over this, and don't expect me to stick my neck out and stop him."

"Yes, sir."

"Wherever you are, turn around and get in here. Now!"

Thirty-two years of trying to keep my relationship with my father on an even keel was hard to overcome, but it occurred to me that trying to handle him now was just going to backfire later. There was no question that he was going to find out about my trip. That was probably the whole point of it.

"I can't. I'm on a plane."

"A plane? Why? Where are you going?"

"I'm on my way to talk to our attorneys in Montana."

"Jesus Christ, Trevor! Those guys are up to their asses in that case right now. They don't have time for one of your stupid history projects. And after what you did today, you should just be sitting in your office, working your butt off and hoping Trainer's got enough on his mind that he forgets about you."

I stiffened a little at that. I doubted my father had any idea what I did for the company and while I had my failings as a human being and a son, I wasn't stupid.

"Paul Trainer asked me to do it," I said.

"What? What did you say?"

"I said Paul Trainer asked me to go to Montana and talk to our attorneys."

"Paul Trainer asked you?"

"Yeah. He said he needed a feel for what's going on up there," I said angrily, getting as close to overtly attacking my father as I ever had.

There was a long silence over the phone.

"You're breaking up," my father said finally, and then the line went dead.

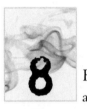**8** By the time I ducked through the jet's door and into Montana's completely transparent air, there was already a car easing to a stop at the base of the steps. A man who seemed to be wearing a suit for the first time in his life burst from the vehicle and skittered around to open the back door for me.

"Can I help you with that?" he asked, a little out of breath from his five-yard sprint.

I clutched the leather briefcase in my hand a little tighter and shook my head gravely. "I'd better hang on to it."

Truthfully, the case was just for show—it contained an unmarked yellow legal pad, a cheap ballpoint pen, and a pear. It gave me something to do with my hands, though, and hopefully made me look more official.

My driver jerked his head in a way that suggested agreement and servitude, then began to run back around the car. He stopped short somewhere near the front bumper, turned, and ran back to close the door behind me. Shortly thereafter, we were off.

"It'll just be a few minutes, sir," he said. "Mr. Stone, Mr. Alexander, and Mr. Reeves are waiting for you at the office."

Despite the late hour, all three lawyers appeared to be freshly scrubbed and pressed. I told myself that it was for my benefit, but it was more likely that people just didn't sweat or wrinkle in this climate. Beyond good grooming, though, they had little in common.

Only one of them moved when I entered the office, striding up to me and taking my hand in a slightly painful, but sincere grip.

"Mr. Barnett. It's good to meet you. I'm Steve Reeves."

The name seemed to exist somewhere in an unused corner of my mind, and it took me a little while to recall that it belonged to an actor who had once played The Hulk. Strangely appropriate. Steve was about forty, with a face that seemed to have seen twice that many years' worth of sun and wind. He had one of those thin frames with slightly outsized shoulders that marked him as an athlete and not just a person who forced himself into the gym three times a week.

"Nice to meet you, too, Steve. Call me Trevor."

He smiled broadly and waved a hand toward the man standing behind him. "Let me introduce you to Frank Stone."

Stone's grip was a bit gentler: rough skin covering puffy flesh. He was probably in his mid-forties, with a serious expression and a faded brown tie that for some reason made me certain he'd lived in Montana his whole life. I'd have given odds that he was a state senator or maybe even the town's mayor.

". . . and I think you know Dan."

I looked down at a man who obviously had no intention of abandoning the bench he was sitting on. We'd met few times before, though I'd never heard anyone call him Dan. His suit was dark gray, not far removed from black, and contrasted violently with his white shirt and bloodred tie. He had a narrow face that seemed to come to a gradual point at the tip of his sharp nose, giving up aesthetics for aerodynamics and making him look a bit like a rat. Actually, that description was probably tainted by what I knew about him, but it gets the point across. I nodded politely, and he gazed back at me through an expression that was undoubtedly calculated to be neutral but came off more as put out.

And why not? Daniel Alexander was a former Harvard Law professor who now coordinated the tobacco industry's more high-profile litigation and reported directly to my father. When we'd met in the past, he'd been dismissive and abrupt—quickly conveying that his time was too important to spend on someone like me. I suspected that his feelings hadn't changed.

"Hello, Daniel," I said, forcing him to acknowledge me. "It's good to see you again."

"Likewise."

We all looked at each for a few seconds.

"What is it we can do for you, Trevor?" Reeves said, finally.

"Paul Trainer sent me here to get a feel for what's going on. I think he feels a little cut off, being all the way across the country and all," I said, sticking to my mental script. "Obviously, this is on top of his priority list right now."

"We've been sending detailed reports," Reeves said, sounding a little worried. "Pretty much daily . . ."

"I know," I said, though I really didn't. In fact, I'd been provided no background at all for this trip. I was working only with what I'd read in the papers and heard through the slightly panicked office grapevine.

"Paul's shaky on long, technical reports," I said. "He'd rather get his information a little more . . . personally."

That solicited a condescending snicker from Alexander. "Paul Trainer reads everything he gets and understands everything he reads."

We all waited for him to elaborate, but it became apparent that he had nothing more to say.

"That may be true," I said, ad-libbing a bit. "But it's hard for him to get a handle on the intangibles from a report . . . The, uh, atmosphere."

Reeves nodded slowly. "I don't know exactly what to tell you, Trevor. In the end, it's going to come down to the jury." He waved toward a large piece of posterboard stuck to the wall. It contained twelve photos, with a few paragraphs beneath each, and a cigarette graphic beneath about half—I assume to indicate that the juror smoked. Stepping forward, I examined the faces in the photos a little more closely.

"It's a fairly even split," Alexander explained, his compulsion to pontificate finally overwhelming his desire to ignore me. "Seven are hard-core locals—mostly people with limited education who've never been fifty miles from this town. Four of them smoke—"

Stone cut in, sounding a little insulted. "The other five moved here in the past few years. They all have college degrees, but don't really do anything with them. Mostly work in coffee shops or some kind of retail. Hippie types. People who'd give their eyeteeth to get pot legalized, but love to bitch and moan about tobacco. Two of them smoke, but not much."

That explained why these two local attorneys were doing most of the heavy lifting in court. They each represented one of the town's factions. The outdoorsy transplants would identify with Reeves, while the people whose families had started ranching this area a hundred years ago would be swayed by Stone. It was a good strategy, since I couldn't see anyone outside the borders of Manhattan warming up to Daniel Alexander.

"On an emotional level, there's a lot for the plaintiffs' attorneys to work with," Reeves said. "The hippies, as Frank calls them, are very susceptible to arguments about suffering caused by cigarettes and the financial burden of health care for the people who get sick. And . . ." He glanced at Stone, smiling good-naturedly. "The rednecks, as I call them, are just as susceptible to arguments that the industry is taking advantage of them and thinks they're stupid. Honestly, our best bet is going to be playing the rednecks and hippies against each other. There's a chance that they might not be able to agree out of simple personal animosity. That would at least get us a hung jury."

"We can't bond off a two-hundred-and-fifty-billion-dollar judgment," I said, stating the obvious. I decided to go ahead and speak the words that no industry official had yet uttered outside the walls of Terra: "If we lose here, it may be the beginning of the end."

The two local attorneys suddenly looked a little ill, obviously unaccustomed to having the future of a multibillion-dollar industry hanging on their moderate legal skills. Daniel Alexander just looked bored.

"Like I said," Reeves started, "I'm not going to lie to you, Trevor. It's an uphill battle. The representatives of the class are really compelling . . ."

Like everyone else in America not in a coma, I was familiar with the three people representing the 500,000 plaintiffs. Two were dead and testifying via prerecorded videotape—something that did a fair amount for their credibility. Both had been upstanding citizens, both had tried to quit on numerous occasions, both had started when they were young, and both had cancers strongly linked to smoking.

"Mrs. Glasco is a major problem for us," Reeves said, referring to the surviving representative.

"Goddamn schoolteacher," Alexander grumbled.

The faces of the two local attorneys darkened for a moment, but Alexander didn't seem to notice.

"Mrs. Glasco is extremely popular in this town," Stone said. "She taught first grade for years and has universal appeal."

"The bottom line here," Reeves said, "is pretty much everyone agrees that the woman's a saint."

"Does that include you?" I said to Stone.

He shrugged. "I had a little dyslexia when I was a kid. She stayed after school every day to help me with my reading."

"What about you, Steve?" I said. "She wasn't your first-grade teacher, right? You haven't lived here your whole life."

He shook his head. "But she taught my two kids."

"And she's milking it like you wouldn't believe," Alexander cut in. "Rolling around in that goddamn wheelchair with the oxygen tank on the back and talking about God all the time. God this, God that. God's coming to take me away. You have to give the plaintiffs' attorneys credit. Unless there were nuns or crippled children available, they couldn't have picked a better representative of their class."

I nodded thoughtfully, trying to maintain the illusion I knew what I was doing. "And the judge?"

"Openly hostile, but a stickler for the law," Alexander said quickly. "Read *Hamilton* v. *Reid* or *Lucas* v. *Dawson* and you'll get a feel for the fact that he doesn't let his personal feelings get in the way. He set aside a jury verdict in Hamilton for legal reasons despite the fact that it was well known he agreed with the verdict personally."

Stone and Reeves looked at him, seemingly a bit perplexed. They probably hadn't bothered to go back and comb through the judge's record to get a feel for his personality. I guess that's why Alexander's was a household name and theirs weren't.

I pulled the pad from my briefcase and jotted down what he had said. I was anxious to go back with at least a few morsels of good news, preferably in sufficient detail to make me look like I had some clue of what I was doing. I had no idea where I stood with Paul Trainer, but it seemed certain that displaying a little competence and diligence couldn't hurt.

"So you're saying that if the jury comes up with a huge award, the judge might overturn it? Then we wouldn't have to worry about an appeals bond?"

"Based on his history, it's possible," Alexander said, starting to warm up to the subject. "Obviously, we're going to try to win it outright, but this is at least some kind of a fallback if the jury goes off the deep end and decides to hand that damn schoolteacher a quarter of a trillion dollars."

9 I PUSHED ASIDE THE REMNANTS OF A PLATE OF BA-con and eggs and immediately started sucking up an order of biscuits and gravy. A few hours ago, I'd claimed a prized table by the windows and I felt obligated to keep eating as long as I was tying it up.

I concentrated on my food, ignoring the overcaffeinated stares of the oddly sophisticated-looking people packed into the diner with me. Homogenously dressed in a style that could only be described as "urban safari," they'd replaced the weathered, baseball-cap-wearing men who'd left an hour ago, and had been waiting to get their hands on my booth ever since.

"I've got a bet with the cook that you can't finish that."

The waitress bent over a little as she filled my coffee cup, and the male patrons of the diner realigned their covetous gazes for a moment.

"Really? I was just about to order a Danish."

She was one of those girls who managed to be even more intriguing in a horrible polyester uniform because it forced you to imagine her without it. A welcome distraction.

"I haven't seen you in here. Are you with the rest of these guys? Are you a reporter, too?"

"Just passing through," I said and she strolled off, looking kind of disappointed.

I shoveled another bite of sodden biscuit in my mouth, chewing lazily, and turning my attention back to my unobstructed view of the courthouse across the street.

Angus Scalia was standing on the top step, speaking with malicious relish to an enthusiastic crowd that he'd undoubtedly bussed in. I couldn't hear him, but I was close enough to see his thick lips distorting and the sweat stains appearing beneath the armpits of his shirt. He turned to the side, revealing his infamous Alfred Hitchcock profile, and jabbed a finger at the building behind him.

The crowd started to applaud, creating a vague hiss that penetrated the window. I leaned in a little closer to the glass, managing to pick out his voice, but unable to understand what he was saying. Probably comparing people involved in the production of cigarettes to Satan. Or Hitler. Or smallpox. Or Jack the Ripper.

Despite his tendency for going a little over the top, it had always been hard for me to dismiss Scalia as the quixotic crackpot my colleagues nervously labeled him as. He was the only antitobacco crusader I knew of who actually had an agenda. I know it sounds cynical, but antismoking lobbyists are perhaps the most ineffectual bunch of yahoos who ever walked the face of the earth. The simple secret to controlling them—and believe me, the tobacco industry did—was making sure they were well funded. Every time Big Tobacco took a hit and had to pay out some money, it "reluctantly" included tens of millions of dollars in installment payments for antismoking efforts. Then we'd slip in a clause that said any reduction in industry profits would reduce these payments. So, easy as pie, we turned the antismoking organizations into self-perpetuating machines with absolutely no real incentive to reduce smoking. Clever? Not really. These guys were just too easy.

Except for Scalia. He resisted every effort to latch him to the tobacco teat and missed no opportunity to denounce his antismoking compatriots for the bunch of clowns they were. As near as I could tell, the man was hated by almost everyone involved: his colleagues, Big Tobacco, the government, smokers. Everybody but the media.

He was an amazing sound-bite guy—a man who could easily be riled up enough to spray spit all over the camera lens and almost always looked like he was inches away from a serious cardiac event. No two ways about it: That just made plain good television. And now, with the Montana suit in full swing, his politically impossible ranting was starting to resonate. He seemed less like he was tilting at windmills and more like a man who, through sheer, unwavering persistence, might turn out to be a winner.

"Uh oh," I mumbled through a mouthful of gravy. They were giving him a microphone.

"I'd like to bring someone up here who needs no introduction . . ." The sound of his voice vibrated the glass next to me and caused the other patrons of the diner to start slapping money down and running for the door.

The old woman being pushed by one of the plaintiffs' lawyers waved weakly as the crowd thickened—not only from the weight of the reporters abandoning the diner, but also from townspeople who happened to be walking by. They all looked up at her with a sad reverence that made my stomach tense.

Steve Reeves had given me Mrs. Glasco's deposition, and I'd skimmed it last night before I'd gone to bed. She was only sixty-five, though chemo had added about twenty years, as it always seemed to. She still lived in a neat but tiny house that was all she'd ever been able to afford. Since her husband died and she'd gotten sick, the local kids took care of her yard and whenever anything went wrong with the house, it just got fixed by one of her former students at no charge. The only conclusion I could come to was that she wasn't human. No one was that well liked. I mean, I'd always hated my teachers.

She just sat quietly, her wheelchair dangerously close to the steps, while Scalia railed against the people who had done this to her—the people who had addicted her at a young age while suppressing information about the dangers of the habit, the people who had tried to negate the government warnings through smooth denials and slick advertising. The people who now fought tooth and nail to shirk their responsibility to her.

The sad thing was that it seemed like that was all there was left of Mrs.

Glasco. She was now just potential income to her attorneys, potential losses to my company, a poster child to the antismoking lobby. And when she died, no one would ever think about her again.

No, that wasn't true. There were still the kids she'd taught—a whole townful of them. They'd remember her as more than a political lightning rod.

The pretty waitress came around again as I dropped a twenty on the table and stood. "I win," she said.

"What?"

"You couldn't finish it."

"Oh yeah. Right." I started for the door but then stopped and turned back toward her. "Did you have Mrs. Glasco?"

"Nah," she said cheerfully. "I had Mrs. Blake."

I JOGGED ACROSS the street, the still-cold air cutting through the front of my dress shirt while the sun burned into the back.

"We've put up with hundreds of years of this," Scalia shouted as I melted into the outer ring of his audience. "Hundreds of years of lies. Hundreds of years of manipulation. Hundreds of years of death. But now we have the power to send a message—to take the first real step toward stopping this murder machine."

Scalia wasn't an elegant speaker—too prone to clichés and melodrama—but those technical failings never seemed to lessen his impact.

"How many have they killed over the years? How much have they cost this country?"

I glanced at my watch and confirmed I was running late. Our lawyers had invited me to observe some of the trial before I went back. Honestly, I didn't see the point, but it might make me seem more diligent and interested if I showed up for a little while.

I skirted around the edges of the crowd while Scalia continued on pretty much the same oratory path. I'd nearly made it to the steps when he fell silent for a moment.

"Look right down here!" he suddenly boomed.

I tried to squeeze between a woman and the baby carriage she was standing next to, but ended up having to take the long way around.

"I want you all to look right here!"

I happened to glance up at just that moment and discovered that he was pointing right at me. I froze with one foot on the steps.

"Do you know who this man is?" he asked the crowd. "He's Trevor Barnett. His family has been one of the most powerful forces in tobacco for almost two hundred years."

I knew I should just lower my head and run for the courthouse doors, but I couldn't seem to move. How did he know who I was? I remembered Stan's paranoid ranting.

They've got files . . .

"The industry must be worried," Scalia said as I stood there like a deer in his headlights. "We must have them scared for them to send one of their own here."

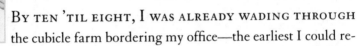

10 By ten 'til eight, I was already wading through the cubicle farm bordering my office—the earliest I could remember ever setting foot in Terra's headquarters. It wasn't intentional: I'd planned to play things cool, since I figured my fate was already pretty much determined. But I'd come wide-awake at six A.M. and every effort to roll over and fall asleep again had failed. Nicotine almost had a heart attack when I'd pounced on her in her dog bed. Paybacks were hell.

About a quarter of the staff was already in, but none were working—mostly they were milling around talking and searching for coffee. I tried to act relaxed, shoving my hands in my pockets and smiling serenely at people who seemed to be fighting an urge to back away.

I guess it was possible that my smile wasn't so serene. I'd managed to keep my mind reasonably clear through a rare display of willpower, but there wasn't anything I could do about the nervous excitement in my stomach. Intellectually, I knew none of this had anything to do with me—that I was just a pawn in a game that was inevitably going to end badly for me. But it was hard not to pretend just a little. I mean, I'd been sent—on Paul Trainer's personal jet—to interpret one of the most important events in the long history of tobacco. It wouldn't hurt to let myself enjoy that for just a little while. In a few days, Trainer would have made his point to my father and he'd can me. There would be plenty of time then to scrape and grovel. There always was.

I could hear rustling coming from Stan's cube and I flopped my forearms across the top of it, laying my chin on them: one of the benefits of being tall in the world of the modern office.

"What's going on, man?"

He seemed a little startled to see me.

"You're in early, Trevor."

By now, everyone on the floor knew about my report to the board and that Paul Trainer had come to my office yesterday. The gossip mill would be working overtime: speculating, polling contacts on other floors, rehashing my history at the company. Did everyone seem skittish because they'd concluded I was the walking dead? Did they think Paul Trainer was going to send a bunch of goons down here to get rid of me and everyone I'd ever known?

"Couldn't sleep. So what's the word, Stan?"

He didn't seem to want to scoot back in his chair, so he had to crane his thick neck to see me. It was like he wanted to look like he was working.

"Nothing interesting, Trevor. It's been busy, you know?"

I laughed. "Yeah, right."

A silence started to spread out between us, and I decided to just let it happen. I didn't have anything better to do.

Finally, Stan got uncomfortable enough to speak again. "I saw you on TV."

For some reason, that hadn't occurred to me. Thinking about it now, it did seem a little unlikely that the media would ignore a face-to-face confrontation between Angus Scalia and the old tobacco money they figured I represented.

"Really?"

"What an asshole," Stan said. "You should have decked him. Knocked him on his fat ass."

And that was it. He didn't ask why I'd been there or what I'd learned about the suit that might or might not put him out of work. He just smiled politely, put his head down, and dug into a stack of papers on his desk.

WHEN I EASED through his office door, Paul Trainer jumped out of his chair and threw an arm around my shoulders. All in all, not an easy maneuver for someone his height and age.

"How was your trip, Trevor? Good? How'd you like the jet?"

"It was fine, sir. Thank you for asking."

He motioned toward three men standing by a silver cart parked in the far corner of his office. Two of the men helping themselves to pastries and coffee were board members I recognized but whose names I couldn't remember. They each gave me a polite nod. My father didn't bother to turn around, instead dumping some cream in his coffee and taking a seat at the table.

"We saw you on TV," Trainer continued energetically. "You looked good. A little slack jawed. But good."

Kind of hard to know how to respond to that, so I didn't.

"Well?" Trainer said, releasing me and sitting down behind the table my father was ignoring me from. The others did the same but I wasn't invited, so I just stood my ground and started into a summary of what I'd done over the last twenty-four hours. Not an easy thing to screw up, no matter how scared you are. I let my eyes wander across the low-hanging portraits on the wall, hoping that I would appear to be confidently scanning my audience and not looking slightly over their heads.

". . . then finally, yesterday, I sat through an entire day of court before I flew back last night."

Trainer was wearing a slightly impatient expression but was making an uncharacteristic effort to hide it.

"And?"

"Based on what I saw, the jury's against us. The plaintiffs' attorneys are doing a good job of painting the industry as evil, and the jury's buying it. The judge seems to be personally against us, too, but I'm not sure that's going to hurt us."

My father surprised me by speaking up. "How so?"

I didn't answer immediately. Was he actually asking my opinion?

"Well, in *Hamilton* v. *Reid* and *Lucas* v. *Dawson* you can see that he doesn't let his personal feelings get in the way. He set aside a jury verdict in *Hamilton* that he personally agreed with." The statement was more or less directly plagiarized from Dan Alexander, but I figured I'd be better off not trying to paraphrase a former Harvard Law professor.

"I'm not sure that's a completely accurate interpretation of the facts," my father said in a sympathetic tone that lacked even a shred of sincerity. "In *Hamilton* v. *Reid* the judge reversed the jury *in favor* of a strongly held personal belief—not against it. It was later struck down by the appellate court." He looked directly at me with a disappointed expression that made me feel like I was twelve again. "I know it can be confusing."

My mind completely locked. I was sure I hadn't confused anything Alexander had said. But somehow I must have. He was one of the country's top attorneys. He didn't make mistakes.

"And in *Lucas* v. *Dawson*," my father continued, "the judge had fairly well-documented political reasons for reversing the verdict. I think everyone here is familiar with them. This case is obviously different. The political benefit accrues from going against us. It also feeds his ego, which is something else he's prone to."

No! I was sure. I'd written Alexander's words exactly as he'd spoken them. "But this came straight from Da—"

"I think you just misunderstood him, Trevor. I'm confident that Daniel understands these two verdicts—they're fairly simple in the scheme of things. He was probably just making a joke and assuming that you'd have reviewed the cases before you went out there."

I felt the now-familiar sensation of blood rushing to my face, and I would have turned my back and walked out if I could have made my feet move.

Trainer finally spoke up, earning him what I thought at the time would be my undying gratitude.

"Then what I'm hearing is that everybody's against us on this. The jury's gonna screw us purely out of spite, and this judge is going to go along for the ride no matter how illegal and emotional the verdict is."

"We haven't lost yet, Paul," my father said.

"I'm starting to wonder, Edwin . . . What do you figure my odds are?"

"It's hard to say. I mean—"

"What are my odds?" Trainer repeated.

I considered backing up and slipping out the door. I hadn't been dismissed, but it seemed like that was just because everyone had forgotten I was there.

"I don't think you can—"

"Jesus Christ, Edwin! It's a simple question! What are my goddamn odds?"

"One chance in five we win," my father finally conceded.

"Thank you! One in five. That's a number I can work with. So what you're telling me is that before long, I and the other CEOs are going to have a two-hundred-and-fifty-billion-dollar judgment that we can't bond off no matter how certain the outcome of the appeal might be. You're telling me that I'm completely at the mercy of twelve cowboys sitting in a jury box in Montana. And you're telling me that there's nothing I can do but bend over and take it."

"This isn't about the plaintiffs," my father said. "It's about the attorneys. And they're going to want their money. They'll settle, toss their clients some spare change, and walk away rich."

"Only a few billion?" Trainer mocked. "And what about the fifty thousand ambulance chasers who are lined up watching what's happening in Montana and licking their chops? We're going to find ourselves fighting class actions from people because they think it's our fault that they can't take a satisfying dump. We're going to spend the next ten years being bled to a slow, lingering death."

I started edging backward as Trainer pulled a cigarette from a pack lying in front of him and lit it. His skinny chest expanded visibly as he pulled on it.

"You know what I feel? I feel pleasure, calm, familiarity . . . Relaxation," he said, then walked over to me and handed me the cigarette. "Tell me something, Trevor. What do you feel?"

I took an unusually long drag, feeling the smoke fill my lungs and the weight of everyone's eyes on me.

To this day, I have no idea why I didn't just paraphrase what Trainer had said. Maybe I wanted to show off for my father. Maybe subconsciously I still wanted that pink slip and the terrifying freedom that would accompany it.

"I feel . . ." My voice faded for a moment. "I feel hyperplasia of the epitheleial cells turning to preinvasive lesions. I feel carcinogens penetrating the nucleuses of my cells and mutating their genetic makeup, I feel my cilia being slowly paralyzed . . ."

The people at the table—my father especially—were so still they looked kind of like the propped-up corpses of Old West bank robbers. Trainer, on the other hand, nodded thoughtfully, relieved me of the cigarette, and wandered back to his seat.

"Trevor's right," he said. "The danger is all anyone thinks about now. They want to have their cake and eat it too. They want all the pleasure and none of the risk."

"That's not right," I heard myself say.

"Trevor, that's enough . . . ," my father cautioned.

Trainer held a hand up, silencing him. "What's not right?"

"They want the risk," I said. "It's part of the image now. They just want to be one of the lucky ones."

Trainer smiled. "And if you turn out not to be one of the lucky ones you want to be able to tell yourself that you're not responsible. You want to spend the last days of your life trying to prove to yourself that it's someone else's fault you're dying." He nodded respectfully in my direction. "I stand corrected, Trevor."

11 DARKNESS HAD BROUGHT COOLER TEMPERATURES for the first time I could remember that summer, and I turned the air-conditioning off in favor of a rolled-down window. I was probably twenty miles from town, bumping down a maze of bad roads cut out of the area's rich farmland. A piece of paper containing an elaborate set of directions was taped to my dashboard, and I was forced to slow at every intersection to search for obscure landmarks.

The empty road came to a T near a collapsed barn and, for no reason, I eased the car to a full stop. The notes I'd taken in my meeting with Daniel Alexander were sitting on the passenger seat, and I suddenly felt a compulsion to read them for a fifth time. I reached out, but then withdrew my hand. There was nothing to find in those pages—I'd written them verbatim. Had my father been right? Had I just confused Alexander's deadpan delivery of an unfunny joke? There was a much more obvious explanation, of course, but I wasn't ready to think about it. Not yet. My world was already getting way too complicated.

I reached for the notes again, this time finding the strength (or weakness, I'm not sure) to throw them out the window and floor the car around the corner. For a while, it felt like they were chasing me.

I swung the car left when I reached an unmarked dirt driveway and followed it to an old farmhouse that had been rescued from returning to the soil by haphazard carpentry and slapped-on paint.

My suspicion that there was no light coming from it was confirmed when I turned off my car and stepped out into the darkness. I stumbled over the uneven ground, cursing quietly, until I finally made it to the relative safety of the porch. The moment my feet hit the loose boards, the front door swung open to reveal the blackness inside.

I stepped across the threshold and watched a shadowy figure close the door behind me then followed cautiously as the figure started for the back of the house.

The sudden glare when the door at the back of the kitchen opened forced me to shade my eyes with my hand. I slipped in quickly and my escort did the same, slamming the door immediately behind us to keep the escaping illumination from giving away our position to the imagined tobacco industry commandos hovering just outside.

"Did you have any trouble finding the place?" John O'Byrne asked, giving my hand a grave, but thorough shaking. He was the founder and director of Smokeless Youth, a small but influential antitobacco lobby group aimed, not surprisingly, at reducing teen smoking.

"None at all, John. Sorry I had to cancel on you before."

"No apology necessary. I understand how difficult your situation is." He pointed to a chair. "Have a seat."

I lowered myself into a rickety chair that seemed barely able to support my weight and O'Byrne did the same, positioning himself directly across a small table that looked like it had been in the room since the house was built.

So why was I sitting down with the enemy? Kind of a long story.

After the industry had agreed to pay hundreds of billions to reimburse the states for their medical costs in return for the promise that they wouldn't sue again, tobacco stocks had enjoyed a substantial increase in value and my trust checks had briefly swelled. Flush with cash but still plagued by a vague sense of guilt that my halfhearted forays into religion, philosophy, and self-help hadn't expunged, I'd given SY a fairly sizeable donation. The theory here was that I'd part with a little of what I'd made from dooming Mrs. Glasco and people like her to cancer, thus tidying my soul and easing my conscience. As it turned out, giving away a bunch of money I hadn't done anything to earn didn't have much of an effect on my soul one way or another. So, in the end, the applicable theory was "easy come, easy go."

Ironically, the industry's seemingly unstoppable lawsuit-induced slide began shortly after my donation and it quickly became apparent that I'd actually given more than I could afford. So from a spiritual standpoint, I'd paid a whole lot for absolutely nothing.

Well, not *absolutely* nothing. I'd unintentionally landed an unpaid position on the board of SY. John, like most antitobacco pundits, liked only one thing better than building his permanent home on the moral high ground, and that was feeling like he was putting something over on the industry. In me, he'd recruited an insider—a spy, if you will—that he could use to . . . Well, to do whatever it was he did.

If I'm such a cynic, why even accept a position on the board? Moreover, why continue to shamelessly maintain the illusion that another big check was right around the corner, even though I was broke? I had my reasons.

"You're sure you weren't followed?" O'Byrne said. "We picked this location because you'd be able to see a tail for miles."

"Positive," I said.

"We saw you on TV," he said, motioning toward his assistant, sitting next to him. "What were you doing in Montana?"

O'Byrne had a somewhat gaunt, excitable face and constantly busy hands that were strangely infectious. He gave everything the same urgency as a child holding a secret meeting in a tree house.

"Paul Trainer asked me to go and evaluate what was going on up there." I was a little embarrassed at the spark of false pride I felt when I spoke.

"Paul Trainer? You're talking directly to Trainer now? When did this happen?"

"Just a few days ago. But I don't think it's a big deal. He just wanted a different perspective."

"Jesus, Trevor! Not a big deal! If you could cultivate a relationship with Paul Trainer, can you imagine what it could mean for us?"

I honestly couldn't.

"So what *is* going on in Montana? What did you find out?"

"I think we're going to lose," I said. "You wouldn't believe what a circus it's turned into."

"I saw you and Scalia face off," O'Byrne said. "It was on every channel." He shook his head slowly, as though Scalia was a crazy brother-in-law who'd knocked over a liquor store. Then he changed the subject. "This could be the turning point, for us, Trevor. Seriously historic stuff. What do you think the companies will do if they lose?"

"I don't know. I guess they'd all bankrupt themselves and then—"

"Of course they will—just more of the same. More repugnant legal maneuvering. But this time they won't get away with it. Do you think the Montana attorneys will hold to the judgment? That they'll go for the whole amount and make the companies sell off their assets and close their doors?"

I shook my head. "I think—"

"No, you're probably right. They'll want to be paid. But they'll get their pound of flesh. They'll get a few billion . . ."

It went on like that for half an hour: John asking me questions and then answering them himself. As always, he talked in the broad, flowery language of the antismoking lobby, but was polite enough to tread carefully around words like *evil* and *murder* in deference to my genealogy. I didn't pay much attention. I nodded, let out a few encouraging grunts, and pretended to look at him when I was really concentrating on his assistant in my peripheral vision. She sat motionless through the entire thing, staring straight ahead as though she could almost see the smoke-free utopia that America's next generation would be born into. Or maybe she just wasn't paying attention either.

So there it (she) was: the reason I'd stayed on at Smokeless Youth after it became obvious that it wasn't going to provide me with any meaningful illusion of absolution. John's assistant, Anne Kimball.

At thirty-one, she was one year my junior and, unlike me, she'd actually accomplished something in that time. She'd graduated with honors from the University of Pennsylvania and gone to Georgetown Law on a scholarship. After graduation, she'd signed on with a D.C. law firm and by all reports had been on her way to doing big things. That promising career, though, had been cut short when she suddenly quit and joined SY for what had to have been an 80 percent pay cut. Her colleagues had tsk-tsked, used words like *burnout* and *nervous breakdown,* then fought like pregnant wolverines to get their hands on her office.

What they hadn't known about Anne is that her mother had died horribly (it was the most popular way) of lung cancer about ten years ago, after smoking a pack and a half a day for nearly her entire life. It seemed likely that it was her memory that had prompted Anne to chuck everything and go to work for John.

"It's all lining up, Trevor. It's all finally moving in our direction."

I nodded automatically and continued to watch Anne as O'Byrne went on. And on.

My best guess was that she was almost a foot shorter than me, though she tended not to let me get close enough to get a firm estimate. Her face was extraordinarily round, contrasted by a straight nose and a set of supernaturally green eyes. Her hair was an unremarkable brown, pulled back in a long ponytail that didn't seem to fit with her wardrobe, which tended toward formlessness and Maoist grays. Despite that less-than-colorful paper description, though, she was the most striking woman in the world. To me, at least.

"Now's when we hit them, Trevor! This is our moment!" O'Byrne reached across the table and grabbed my forearm, forcing me to refocus on him.

"The Montana suit, the new surgeon general's report . . . The industry's as weak as it's been in a hundred years. It's time."

"Time for what, John?"

"Time to hit them! And hit them hard."

I raised my eyebrows, prompting him to give me a little more detail on just how he planned to deliver this deathblow.

"The surgeon general's report says that five million teens are smoking. Five million! It's a huge number with enormous impact, Trevor. It's a number that'll wake people up—particularly with everything that's happening."

I chewed the nail of my index finger for a moment. "Actually, John, didn't the report say that five million teens smoked one cigarette in the last month? That's doesn't really make them smokers . . ."

"Who cares? Are you trying to say that the tobacco industry's never exaggerated or told half-truths to drive their point home? We're going to show them that we've got the guts to play their game."

I glanced over and saw Anne rolling her eyes. When she caught me looking, she quickly redirected her gaze to the floor and conjured up an expression of steely determination.

She was one of those rare people who sincerely believed in what she was doing. I had no doubt that when she'd signed on at SY, she'd figured she could use her limitless intelligence, creativity, and passion to slowly dismember the industry that killed her mother. It was hard not to wonder how long it took before reality set in. How long before she realized she'd be spending half her time managing John O'Byrne's ego and the other half groveling for donations that would be pissed away on showy but ultimately pointless antismoking initiatives. Life had a funny way of taking good intentions and ramming them back down your throat.

"What do you figure the industry will do when we roll out with a campaign based on that number?" O'Byrne asked. "How hard do you think they'll come after us?"

What he wanted to hear was "Pretty hard." He was hoping for a backlash weak enough that it wouldn't really hurt him but strong enough that it would be obvious to the world that he, John Samuel O'Byrne, had inserted a thorn into the lion's paw.

And I reckoned that's exactly what he'd get. The companies would draft a tepid protest and mail it off, making sure it didn't have anything in it that might scare him off. You see, nothing made the tobacco industry happier than when the hapless antismoking lobby exaggerated the number of teen smokers. Basically, the message they were sending was: *Come on, everybody's doing it.* A powerful statement when aimed at an age group that puts the importance of fitting in just ahead of food and air.

"They'll come after you," I said, sounding a little apathetic, even to myself. "But it won't be anything you can't handle."

John leaned back into his chair and folded his arms across his chest. "This campaign is coming at just the right time, Trevor. We're really going to do something here."

"Yeah?"

"This could be it," he said again. "This could be a turning point."

I had no idea what this "it" he kept bringing up was. And if this was a turning point, what direction would we end up in? The fundamental problem with the antismoking lobby was that they had no real goal. Compare their ethereal moralizing with the very clear and quantifiable goals of their opponents. I can tell you with complete certainty that it was Terra's holy quest to hopelessly addict every man, woman, and child on the planet to cigarettes.

Not that this was as sinister as it sounded. They were a company in the business of selling product. I assume that McDonald's wanted to hopelessly addict every man, woman, and child on the planet to Big Macs and Jeep wanted to hopelessly addict every man, woman, and child on the planet to SUVs.

"What do *you* think Trevor?" Anne said, looking directly at me for the first time that night. "The Montana suit, the surgeon general's report, this campaign. Is this a turning point?"

She caught me a little off guard. Generally, all I had to do in these meetings was daydream about her and occasionally nod. "Uh, sure. I guess—"

"And you could be a cornerstone to all this," O'Byrne interjected. "With you close to Trainer, we've finally got a man inside . . ."

THE MEETING had been over for fifteen minutes, but I hadn't gone home yet. I was smoking a cigarette beneath the broad branches of a magnolia tree just to the side of the house's overgrown driveway. O'Byrne had driven away in his respectably worn Jetta after shaking my hand with the energy of a man on the hunt for a donation. Anne was still inside cleaning up.

It was another five minutes before she appeared on the porch, juggling a few files while she tried to pull the door closed behind her. My heart started to beat a little faster as I ground out my cigarette and stepped from the moon shadow of the tree. She didn't come to a dead halt when she saw me, but she did slow to a more uncertain pace as she continued toward her car. I matched her speed and we met at the driver's side door. She didn't take into account that my eyes were better adjusted to the darkness and didn't immediately hide the trace of fear on her face. Suddenly I felt like a stalker.

"I admire your effort, Anne—your guts for attacking," I said. Despite twenty minutes of standing around, that was the best opening line I'd been able to compose.

"Yeah, well I guess we'll have the admiration of the whole tobacco industry before long," she said, concentrating on opening the door to her much more than respectably dilapidated Pontiac something or other. "When should I expect the roses from your marketing department?"

"I don't have a marketing department, Anne. I—"

She turned to face me, and I had to fight an urge to step back. A few hundred years ago, those eyes probably would have gotten her burned at the stake.

"So how's it feel to be a cornerstone, Trevor?" The frustration was clearly audible in her voice.

"A what?"

"You heard John. You're going to be a big part of this going forward. Quite an opportunity for you."

When I'd first started on the board, my relationship with Anne had been pretty good. Really good, actually. But it hadn't taken her long to see my involvement in SY for what it was—a cheap attempt to absolve my sins. And I think it made her angry to see me dabbling pointlessly in something that had become the focus of her life.

"I guess . . ."

She tossed her files into the open car and then looked up at me again. When she spoke this time, her tone had moderated. In less than a minute, we'd gone from fear, to anger, to professionalism. The last was the worst by far.

"I'm sorry if I seem rude, Trevor. I guess I'm just a little tired."

At least she didn't call me Mr. Barnett.

"No problem."

She squinted at me through the darkness as though she was searching for something. When she didn't find it, she climbed into her car and started the motor.

My original plan was to ask her out to dinner. For some reason, recent events had given me a new and unfamiliar measure of resolve. I didn't understand the effect and I didn't know how long it would last, so I figured I'd better use it quick.

Now having completely abandoned that plan, I stepped back and let her throw her car into gear. She cranked up the stereo and I listened to Ian Kingwell's three-pack-a-day voice and grinding guitars as she pulled out into the road.

 **12** FOR THE SECOND TIME THAT WEEK, I WAS STARING at my ceiling by six A.M.

My world, which until recently hadn't provided me with much that I couldn't avoid contemplating, was quickly becoming completely unfathomable. How to interpret the intricate plot Paul Trainer was hatching to completely destroy my life, my worsening relationship with my father, my mortifying performance in the boardroom, my nonexistent relationship with Anne?

My plan to free myself from all this seemed to be backfiring and I had no one to blame but myself. Another half-assed execution of a perfectly workable plan in a life already brimming with those kinds of gaffes.

As I lay there waiting for Nicotine to launch her morning attack, I tried to convince myself to blow off work and use the time to write my resume. I was still afraid of the world out there but, for the first time, that wasn't the reason I knew that I was going to get up, get dressed, and make the drive to my office.

Nothing had really ever gone very wrong in my life. Similarly, nothing had ever gone very right. Despite a few rather depressing moments over the last few days, I had to admit that they'd been kind of . . . exciting. It was hard not to feel a little like a child—fascinated by everything and desperate to be accepted by the grown-ups. I was back to the good old days when poisonous cleaning products and wall outlets glowed hot with the promise of untold adventure.

"'Morning," I said. "What's going on?"

The people laying siege to Stan's cubicle tried with moderate success to keep the intimidation, curiosity, and worry from registering on their faces. Within thirty seconds, everyone had remembered a critical meeting, phone call, or memo that had to be dealt with right away. I watched them hurry off in the general direction of the coffee room.

My list of friends was dwindling fast, and I felt the impact of that more than I thought I would. It occurred to me how little it would take for me to be completely alone in the world. I'd never really inventoried and categorized my friends and family before and wouldn't recommend it as a fun way to spend an afternoon. The outcome can be kind of startling.

I leaned over the wall of Stan's cube and gave him my widest, friendliest smile. "So? What's happinin', man?"

"Not much."

I'd swear he was better dressed than usual. His shirt was pressed and his tie seemed almost stylish.

"Oh, come on. Throw me a bone," I said, paraphrasing what he'd been saying to me every day for years. "You're killing me, here."

He remained silent.

"Did someone discover his girlfriend having a lesbian affair with his mother and shoot up a post office?" I prompted.

He shook his head.

"Did some guy get drunk and run his car into the bleachers during a high-school football game, then open up with an assault rifle?"

That elicited the beginnings of a smile. "I wish." He folded his arms across his thick chest, displaying larger-than-normal sweat stains under his arms.

"Truth is, Trevor, *you're* going on. You're getting kind of famous."

"Infamous," I said. Word of my meltdown in the board meeting had undoubtedly oozed down to this floor by now.

"I don't know," Stan said, now crossing his legs in an impressive display of paranoid body language. "Everyone's wondering if they should start calling you sir."

He was testing the waters; handling me. After nine years of doing everything I could to fit in here, I was being handled by my best friend at the company.

"Jesus, Stan. I went to Montana to put some historical perspective on the trial, made an ass out of myself on TV, and then made an even worse ass out of myself in the boardroom. I don't think they're going to give me a corner office anytime soon."

He managed to grin, but it looked like it took some effort. "Sure, man. I know."

And after that fairly ambiguous statement, he just sat there with his arms and legs in knots, wishing me away. The clear message was that he had enough on his plate with the antitobacco lobby, the government, the media, and the court system. He didn't need to deal with a spy.

I stood there for longer than I should have, finally disengaging myself from the top of his cubicle and walking, head down, to my office.

"Chris Carmen wants to see you," Ms. Davenport said as I sulked by.

"When?"

"He said whenever you have a minute."

I sighed and pushed through the door to my office. Chris's ability to call me into a meeting was one of the few unequivocal powers he had over me, and he enjoyed using it. Now it was "whenever you have a minute."

I leafed through my mail without much interest until I came upon a package from Daruis's company. I tore it opened and found a CD with a sticky note on it that said simply, "You're gonna like this one."

I was about to toss it aside, but ended up sticking it in my computer instead. A few moments later the beta version of Darius's latest game was filling my screen. I'm not sure what I was trying to prove, but whatever it was, I started the game and turned the volume up so it would be audible over a good half of the floor.

The first level wasn't too difficult, despite the fact that I was born a few years too late to have split-second thumb timing. After about five minutes of mutilating astro-zombies with a Gatling gun, I manage to procure a key that looked like a flaming skull and my character advanced to the next level.

I sat back for a moment, admiring the movielike realism as the muscular, scarred warrior trudged into a futuristic freight elevator. I was wondering what it was about this particular game that Darius thought I'd like. It seemed pretty similar to all the other ones he'd produced.

Then the animated soldier pulled a pack of smokes from his pocket and lit one. He took a tired, dramatic drag and stared down at the floor with all the fatigue, fear, and pain that living in a world populated with astro-zombies would likely cause. When the elevator doors opened onto a new world full of new dangers, he tossed the half-smoked butt on the ground and crushed it with a spiked boot.

I realized my mouth was hanging open and managed to close it, but other than that, I just sat there frozen. Had Darius included that for me? As a joke? Because my trust wasn't paying out and he was trying to help hook a few kids so I could get a new car?

The screen flickered and the astro-zombies attacked, but my concentration was blown. I tapped the Fire button and maneuvered halfheartedly, managing to hold out for only a few minutes before my opponents began feasting on my flesh in earnest.

"Doesn't that make you dizzy?"

I spun around in my chair and found Paul Trainer standing right behind me. I instantly felt like a twelve-year-old caught extracting a *Playboy* from beneath his mattress.

Before I knew what was happening, I heard myself speaking aloud the words on Darius's note to me. "I think you're gonna like this, Mr. Trainer."

It was like the stories you hear from people who die and are brought back to life: *I was outside my body, floating . . .*

I watched myself jump up and offer Trainer my seat, watched myself show him how to work the gun and how to maneuver the character, and watched myself restart the game.

For a man who grew up before the invention of the wheel, Trainer did pretty well. I occasionally had to jump in and save him with a timely jab of this key or that, but overall he threw himself into the war against flesh-eating ghouls with real enthusiasm.

After winning his desperate battle with the king zombie on level one, Trainer watched his character trudge into the elevator and dig through his pockets for a cigarette. By the time it was lit, the initial shock of being crept up on by one of the world's most powerful industrialists had worn off and I'd had a few moments to reflect on what a truly pathetic human being I was.

"Did you do this, Trevor?"

"No!" I said, trying to regain a little of my dignity. "I had nothing to do with it."

He smiled, and I realized that had been exactly the right (wrong?) answer. This kind of product placement was illegal, and he assumed I was just playing the denial game that had become so much a part of the tobacco industry.

He picked up the sleeve the CD came in, examining the prototype graphics printed on it. "How many of these things ship?"

"I have no idea," I said, backing away from him.

"Like hell you don't," he said, grinning conspiratorially.

"Really! I don't."

"Jesus, Trevor. I'm not going to hold you to the number. Ten? A hundred? A thousand?"

I tried not to answer, but wasn't able to hold out for long. "More. The guy who owns that company says if one of his games only made as much as the movie *Jurassic Park,* he'd consider it a complete flop."

"No shit," Trainer said. "Any way you could get me one? I've had a computer on my desk for years and never found a good use for the damn thing."

I popped the CD out of my computer and slid it into the sleeve in his hands. "It's yours."

"Thanks," he said, as I shifted my weight from foot to foot and tried to decide whether I was more comfortable with my hands in my pockets or out.

"Uh, is there something I can do for you, Mr. Trainer?"

He seemed to have to think about that for a moment before he remembered. "Oh, right. I was wondering if you were going to be at your father's party this weekend?"

I was vaguely aware that my father was having his annual outdoor extravaganza on Saturday, but I hadn't been invited in years. In fact, I hadn't so much as been asked over to the house for dinner in probably a good eighteen months.

"You know, I'm not, Mr. Trainer. I've . . . I've got other plans."

"Cancel them," he said, standing and starting for the door.

"I don't think—"

He stopped suddenly in front of an old poster hanging on my wall and ran his fingers over the yellowing paper. "I remember this campaign . . ."

The extremely rare poster was from World War II, recalling a time when green dye was desperately needed to create camouflage. It depicted a newly designed pack of cigarettes that was almost completely white against the background of a speeding tank. Across the bottom was a slogan: LUCKY STRIKE GREEN HAS GONE TO WAR!

"I was . . . oh, I don't know, nine? I used to help my dad after school. He delivered cigarettes to the stores around Greensboro. I remember how proud I felt."

Perhaps appropriately, the real reason for the pack's color change was the fact that focus groups suggested the green layout didn't appeal to women—a substantial growth market at the time. America's war effort was just a convenient (and admittedly brilliant) excuse for the change. It turned out to be one of the most successful ad campaigns in history.

"Yeah . . . ," Trainer said, seeming a bit lost in himself as he wandered out of my office and into the complete silence beyond. "I felt ten feet tall."

13 I DIDN'T PULL INTO MY GARAGE, INSTEAD COMING to a jerky stop in my driveway and jumping out into the undiminished early evening heat. The subdivision I lived in wasn't new, and therefore didn't suffer from that cramped monotony the modern world seemed so fond of. Most of my neighbors had lived there for at least a quarter century, leaving the area depressingly quiet and devoid of the children that should have been playing in sprinklers, riding down the sidewalk on skateboards, and lamenting the inevitable start of school.

I'd put my house on the market three months ago after finally admitting that tobacco stocks weren't likely to rebound, but no one was interested. Hopefully, the ten grand I knocked off the price would get things moving. According to my calculations, I was going to have to sell my car in a couple of months to keep up with the mortgage payments.

I straightened my FOR SALE sign and trudged toward the house, stopping on my porch and examining the knife stuck in my front door. It was pinning a photograph that depicted two girls in bikinis standing in front of a star-shaped pool. Next to them was a large, studly looking Great Dane, obviously meant for Nicotine.

I pulled the dagger out and went inside, flipping the photo over and reading the back.

TO: TREVOR

FROM: DARIUS

DATE: TONIGHT

RE: YOUNG GIRLS, DRUGS, AND BOOZE

LOCATION: SINSIMIAN (Darius's not-as-clever-as-he-thought name for his party house)

COMMENTS: I TOLD THEM YOU'RE A PEDIATRICIAN WHO JUST RETURNED FROM TWO YEARS IN CONGO DODGING COMMUNIST GORRILLAS (his spelling, not mine) AND FEEDING STARVING KIDS

PEACE BE WITH YOU, MY SON.

Darius had loved what he called "dramatic message vectors" ever since high school. The sad thing was that now the job of creating and delivering these invitations had been delegated to his already overworked assistant.

Nicotine didn't rush forward to greet me, instead watching with a slight tilt to her bushy white head as I kicked the door closed. It was no mystery why. I'd been leaving her alone too much over the last few days and even when I was here, I hadn't been paying attention to her. I started to feel a pang of guilt and for some reason it made me angry.

"What are you looking at?" I said. "I'm out there all day, working my ass off so you can live the good life. Yeah, you're feeling pretty sanctimonious now, but what happens when you have to eat the generic-brand dog biscuits?"

She lowered her head and slinked off.

"That's what I thought," I called after her, deciding not to dwell on the fact that I was yelling at a dog for being judgmental.

I hit the Play button on my machine and walked into the kitchen listening to the authoritative voice of my father vibrate the walls.

"Trevor—I'm having a party tomorrow. It starts around noon. I hope to see you there."

I leaned back out the door of the kitchen and stared for a moment at the machine. What was that all about? Had Paul Trainer talked to him?

No, that wouldn't be his style—he wasn't a permission asker by nature. Maybe the invitation was Dad's idea of an apology for making me out to be a complete jerk in the boardroom? Didn't seem likely.

I slipped back in into the kitchen and tossed Darius's invitation into a gigantic stainless-steel trash can centered in the floor, then pulled a beer from my oversized refrigerator. I'd spent fifty thousand dollars making my kitchen into the epitome of modern efficiency—a place where I could exercise my fascination with cuisine and pretend to be a famous chef. The rest of the house still looked like a partially furnished version of the one the Brady Bunch lived in, and according to my realtor, that had something to do with the fact that it hadn't sold.

I heard Nicotine's claws tap on the floor behind me and I turned, crouching as she approached. "Sorry, girl. It's been a tough week, you know? How about a movie? Just you and me?" I retrieved a DVD hidden in the waistband of my slacks and she grasped it gently in her jaws, growling playfully.

"Nine o'clock sharp," I told her as she padded toward the living room.

THE AREA of town I was driving through wasn't bad in the sense that I thought I was going to get carjacked; it just had kind of a tired shabbiness to it. Peeling homes that had been converted to apartment buildings sat on small, poorly maintained lots and mixed randomly with cube-shaped buildings advertising services like upholstery and vacuum cleaner repair. House numbers weren't plentiful, but there were enough visible that with a little creative math I was able to calculate roughly where I was. When I figured I was within a few blocks of my destination, I parked next to the curb and continued my search on foot.

It took about another ten minutes to find the building I was looking for—a white rectangle with space-age architectural embellishments that suggested it was about forty years old. I squeezed through the cars parked in front and jogged up a set of metal stairs, working off some of my nervous energy before coming face-to-face with number 202.

I'd promised myself that I wouldn't just stand there paralyzed and I kept that promise, giving the door an immediate authoritarian knock. It was possible that some of my courage was the result of being sure no one would be there at nine o'clock on a Friday night, but I chose to ignore that possibility.

When there was still no sound from inside the apartment after ten seconds, I relaxed and started for the stairs. I'd only made it a few steps before the door opened.

"Trevor?"

I turned casually, as though it was just a chance meeting, and smiled. She was wearing a long shirt and a pair of sweatpants. The pink socks on her feet explained why I hadn't heard her walking on the other side of her flimsy front door.

"Anne. Hi. How are you?"

Through subtle questioning of her boss, I knew that Anne didn't have a boyfriend. I assumed that this was simply because she couldn't choose among all the handsome, muscular, suave, and intelligent men in love with her.

"How did you know where I live?"

"Uh, well, when I joined the board they gave me a copy of the phone list."

She considered my explanation for a moment and seemed to find it plausible.

"Why are you here?"

"I wanted to ask a favor. May I come in?"

"Is that the favor?"

If I'd been smart, I would have thrown myself over the railing and hit the ground running. But that probably would have made our next meeting even more awkward than it was going to be already.

"No. I was hoping to ask you the favor while I was inside."

She looked at me like I was there to seduce her into smoking her first cigarette. *Come on, five million kids can't be wrong.*

"Um, I'm not sure that's such a good idea."

"Why?"

Undoubtedly she had perfectly good reasons, but they must have all been too unflattering to say to a guy who helped pay her salary. She stepped out of the way, but didn't actually invite me in. I figured that was the best I could hope for.

I tried not to be too obvious as I wandered around her apartment and made the most of my opportunity to snoop into her life outside of work. She'd painted the walls bright blue and through an open door at the back of the apartment, I could see an equally startling red surrounding her unmade bed. The little kitchenette packed into a corner of the living area looked impossible to cook in, but was neat and cheerful. The furniture was strictly Ikea issue, except for an old couch with a quilt on the back that had the threadbare and outdated look of a family heirloom. Next to it on the coffee table was a half-full glass of wine and a copy of *Lonesome Dove*.

It was an incredibly warm, comfortable space, and I would have said so but I'd learned long ago not to comment on people's homes. They saw it as patronizing—instantly comparing it to the imagined grandeur of my inherited palace.

"Sorry about the apartment," she said, not turning to face me until she'd positioned the sofa between us. "It must be pretty sad compared to your place."

And sometimes I didn't have to say anything.

"Could I have a glass of water?"

I watched her hair—free for the first time as far as I knew—sway across her back as she walked to the sink.

"Seems like there should be some kind of government program to help lawyers who use their powers for good instead of evil," I said.

"I don't need welfare."

This was going well.

"You wanted a favor?" she said, handing me a glass that was not so much Ikea as Pottery Barn. "Did you need some information . . . ?"

"Not really, no."

I took a sip of the water, wetting my quickly drying mouth.

"You know, I was about to make dinner, so . . ."

"Do you need some help?" I said, a little too hopefully. "I'm a pretty good chef. In fact, it's probably what I do better than anything . . ."

Of course, I knew she'd say no, but it was worth a try—nothing calmed me like the weight of a pan in my hand and the smell of simmering spices.

"No, that's okay. Thanks, though."

I wanted to take a moment to glance down at the hint of her body through her baggy clothes, but the way she was staring at me—like she was going to explode with discomfort at any moment—suggested I'd better get to the point. "My father's having a party tomorrow. And I need someone to go with me."

After coming to an initial decision not to go to the party, it had quickly occurred to me that skipping it would be crazy. I still knew that I had to get out of Terra and that I needed to burn enough bridges behind me that I couldn't easily return. No point going up in the flames, though. Right?

Still, the thought of having to go stand around and make small talk beneath the disapproving gaze of my father seemed less than appealing. I needed an ally.

"You . . . Uh, you . . ."

When I'd popped out from beneath that tree a few nights ago, I'd surprised her but I think this is the first time I'd really caught her off guard. She recovered quickly.

"I'm having drinks with some friends tomorrow night."

It was obvious to both of us that she was lying but after years of building up to this moment, I didn't have the good manners to just walk away.

"No problem—the party starts at noon. My dad's place isn't far from here, and there's no way we'd be able to stand it for more than a few hours anyway."

There was a moment of strangely compelling confusion on her part before she took the more familiar position of examining the floor.

"Look, Trevor . . . I know you do a lot for the organization. And I appreciate that. And I want to apologize again if I came off as mean last night. I'm sure you're a good guy and everything, but—"

"The party's guest list is going to be pretty much a who's who of the landed gentry of tobacco. I'd imagine that the top executives from all the companies will be there, not to mention politicians and lawyers and lobbyists. Ever meet an actual tobacco executive?"

"Just you."

I couldn't help laughing at that. "I'm not really typical. I'll bet Senator Randal will be there. He almost never passes up an opportunity to suck up to tobacco money."

"I'm a little . . ." She seemed to lose her train of thought for a moment and surprised me by walking around the sofa and sitting down. "I'm not sure what you want from me."

I looked down at her but ended up staring at a framed photograph on the coffee table instead. A woman whose hair seemed a little too solid and tall to have ever been stylish anywhere but in the South stared back with an uncomfortable grin on her face. For some reason, I couldn't break away. Anne's dead mother seemed to be trying to communicate with me.

So now that you killed me, you want to have your way with my daughter. Is that it?

"Trevor?"

I managed to turn back toward Anne, blinking hard. "I'm sorry. Did you say something?"

"I said that I'm not sure what you want from me. Do you want me to help you make nice with these people? You may not have noticed, but I'm not a big fan of tobacco."

"I didn't say you had to be nice. Look, Anne. Things have been really weird for me at work lately, and I have no idea why I'm even being invited. My father and I . . . well, let's just say we don't get along that well."

She opened her mouth, undoubtedly to offer some rote sympathy, but I just kept talking. "So it would be nice to go with someone whose agenda is fairly clear. No matter what it is."

She considered that for a few moments.

"I don't think so, Trevor."

"Why don't you think of it as spying?" I said. "An uncensored glimpse into Big Tobacco. How many antismoking advocates do you think have ever been to one of these parties?"

I had to make this work. I knew myself well enough to know I'd never come back here again if she turned me down.

"I . . . I don't have anything to wear."

The Armani gambit. In my limited experience, this is generally a woman's last line of defense.

"Buy something."

"I can't afford to."

"It seems to me that this would be a reasonable business expense. You can tell John I signed off on it."

I felt a surge of excitement when, after five seconds, she hadn't offered another excuse.

"So, say around eleven-thirty?"

 **14** *"Hmmmm."*

"What?"

"I guess I kind of expected a lawn boy."

I grinned and accelerated along the never-ending dirt road that served as my father's driveway. The oak trees that lined it—not much more than sickly sticks in the blurry old photos my great-grandfather had taken with his newly invented camera—now towered fifty feet on either side. The sun flashed through them intermittently, bathing Anne in a constant flow of uneven light.

She was leaning partway out the window, trying not to get too comfortable in the deep leather passenger seat. Her dress was off-white with big red polka dots, and while she was beautiful in it, it had the look of something borrowed. Obviously, she hadn't taken me up on my offer to smooth the way for the abuse of her expense account.

The most interesting thing, though, was her legs. I don't mean this to sound vulgar but with her so intent on great-grandpa's landscaping, I could stare at them as much as I wanted to—as long as I didn't crash the car. The clothes she wore to work, whether purposely or accidentally, hid the subtleties of her figure and left no real impression other than thinness. The surprising truth was that she was perfect. Long, elegant legs covered in smooth, tan skin. Small, round br— Well, you get the point.

"How many acres does your father have? It seems like it goes on forever."

"I don't know. I never really thought about it."

She wasn't wearing any makeup and she had this tiny chip in her front tooth that was a few shades too white, giving her a healthy, genuine quality that the Barbie doll southern girls I'd dated in the past would have been horrified by. I watched her hair, free again today, blow around in a hair-spray-less frenzy as she leaned a little farther out the window.

I smiled a little wider, unable to help myself. Seeing her there, framed by the old trees and flowering hedges, smelling her shampoo on the warm air . . . I felt an unfamiliar, but not unpleasant, sensation in my stomach accompanied by the sudden realization that I loved her. Was that a noble sentiment? Pathetic? Stupid? All of the above?

In the end, the moment was too perfect to analyze, so I just kept grinning.

Anne finally ducked back through the window and began extracting hair from her mouth and eyes, trying unsuccessfully to put some order to it. "What are you so happy about? I thought you said you hated these things."

I shrugged. "The sun is shining, the birds are singing . . ."

"And all this will be yours someday."

My smile faded and the breath slowly went out of me.

To be perfectly honest, her analysis was more or less correct. In fact, my father didn't really even own this property—it was held in the family trust. When he died, I would become the primary beneficiary (assuming I was still in the employ of the industry) and would get use of it until I passed it on to the children it seemed I would never have.

We came around a sweeping bend, and a long lines of cars parked on the sides of the road became visible. I sped up a little, suddenly anxious to get Anne to the party. A few hours with the Barnetts was enough to convince just about anyone that my life wasn't as charmed as it appeared.

"How's it going, Jimmy? I haven't seen you forever. What've you been up to?" I said, stepping out of the car and handing him the keys.

"Just parkin' cars, man."

I'd known him even longer than I'd known Darius. When we were kids, we used to smoke stolen cigarettes out behind the guest house until my father found out and didn't care.

"Jimmy, this is Anne," I said.

"Hey. Good to meet you." Her smile and handshake had an easygoing friendliness to them that I hadn't seen in a long time.

"Jimmy's dad is the caretaker here. We've been friends since we were kids," I said, conscious of the fact that I was showing off—proving that despite my position as murderer of millions, I had old friends just like normal people.

"Cool." She glanced up the road toward my dad's seascape of a lawn. "This way?"

Jimmy nodded and she started off up the road, looking a little unsteady in shoes that I guessed were borrowed, too.

"Cute chick," Jimmy said. "Seems nice."

"She hates me," I said.

He appraised me in a kind of disinterested way and then slipped behind the wheel of my car. "What's to hate?"

IT WAS ANOTHER wonderfully executed party. In my limited experience, the weather always seemed to cooperate for these things, and today was no exception. The breeze that flowed across the property wasn't as hot and wet as normal, and it was gusting just gently enough to not disturb napkins and hats. The lawn was dotted with open-sided tents shading teak furniture and women whose plastic surgeons had cautioned them to keep out of direct sunlight. Flower arrangements were everywhere— hanging from low-lying branches, in trellises that had been installed just for the party, and in terra-cotta pots so large that they must have been brought in by crane.

The crowd of a hundred or so people was pleasantly spread out, collected in conversational knots of three or four except where they'd bunched up at the food tables. Most were holding lit cigarettes, but only half were actually puffing on them. The smoke was just thick enough to keep the bugs away.

I spotted Anne just ahead, weaving slowly through the guests with her hands clasped behind her back, examining each face with the intensity of someone studying paintings in a museum. I adjusted my trajectory to intercept her, trying to ignore the attention my presence was getting. I'd almost caught her when my father appeared from one of the tents.

"Trevor!" he said, walking purposefully toward me, hand outstretched. His voice was a little too loud and his manner a little too gregarious, as though he were an actor in a theater with bad acoustics.

"I'm so glad you could make it."

My granddad had taken a certain amount of pleasure in referring to my father as the "runt of the litter." Not a terribly flattering description but fairly accurate, I suppose. Dad was a soft, five-foot-ten man with dark, thinning hair in a family of Vikings. Even at seventy-four when he'd died (staph infection, not cancer), Grandad was an inch taller than me. My two uncles had both been even taller, but both had died young (car and hunting accidents).

Grandad had been absolutely devastated by the loss of his big, good-looking boys. My father, with his competition gone, had tried to step in and be the devoted son but it hadn't worked. No matter how much he tried to help, no matter how good his grades, no matter how meteoric his rise at Terra, Grandad couldn't put the deaths of his blond football stars behind him.

"How's work treating you, Trevor? I guess we're keeping you pretty busy."

"Pretty busy," I said, trying to match the volume of his voice for some reason. "But not as busy as you, I guess."

It went on like that for about three minutes—a fairly long time to avoid awkward silences and still manage not to really say anything. Over his shoulder I could see Anne watching us with the same detachment she had the other guests. My father suddenly cut short what could only be called our verbal exchange, slapped me on the back, told me to get something to drink, spun, and disappeared. A fairly obvious clue as to what was coming next.

"Honey!"

I turned slowly, smiling and bobbing my head. "Hi, Mom."

"What are you doing here, sweetie? I'm so happy to see you."

She blinked her wet eyes and gave me a loose hug, keeping her head pulled back far enough that there was no danger of smearing her liberally applied makeup. She released me after a few seconds, and I led her over to where Anne was standing.

"Mom, I'd like to introduce you to Anne."

They shook hands silently, my mother looking a bit perplexed. She was undoubtedly thinking that with a little hair color, a perm, a Wonderbra, and a fluffier dress, Anne would be almost presentable.

"It's so good to see you again," Mom said to her and then turned back to me. "Now, don't ya'll leave until we've had a chance to chat."

And then she was gone.

"I don't remember ever meeting your mother before," Anne said when Mom had retreated out of earshot.

I quoted a T-shirt Darius had printed a few years back to commemorate a party that featured two pounds of top-notch psychedelic mushrooms. "You weren't there, but I saw you anyway."

She opened her mouth to say something, but I cut her off.

"Don't you think we need a drink? I think we need a drink."

"Okay."

The other guests seemed less confused by my presence now, and I got smiles and nods as we worked our way toward a table covered with liquor bottles. The man standing behind it was black, as were the people whisking empty dishes and glasses from the tables, offering hors d'oeuvres from silver trays, and adjusting umbrellas to provide just the right angle of shade. My father would tell you that he liked to do his part for affirmative action, but I suspected that it just reminded him of a time he wished he'd been born into.

"Beer, please," I said. "Anne?"

"That's fine."

The tops were just being popped off when I felt a powerful hand clamp down on my shoulder.

"I saw you on TV. You should have walked up and smacked that prima donna."

"Anne, have you ever met Congressman Sweeny?"

"I don't think I have."

"What do *you* think, Anne," Sweeny said as they shook hands. "Should he have decked ol' Scalia right there on the courtroom steps?"

Anne didn't answer, instead concentrating on Sweeny as he pulled a cigarette from his pocket and put it between his lips. She finally spoke up when a gold lighter appeared in his hand.

"Would you mind not smoking?"

He grinned around the cigarette and his hand rose to his face, but then he realized that she might not be joking. He hesitated for a moment and then excused himself.

Anne looked over at me, arms crossed in front of her, waiting to be scolded.

"There's always shrimp at these things," I said. "Have you seen any?"

WE SPENT the next half hour making the rounds and exchanging pleasantries with various friends of the family, southern politicians, and tobacco industry dignitaries. Occasionally, I'd take part in some "tsk-tsking" about the Montana suit and entertain some pontificating about the antitobacco lobby, but Anne pretty much kept quiet. I wasn't sure if she was feeling guilty about Congressman Sweeny (a card-carrying scumbag who had undoubtedly been trying to figure out how he could lure her into a broom closet for a quickie), whether she was playing the role of quiet spy, or whether she was just trying to be on good behavior in front of a Smokeless Youth board member.

I noticed her drumming her fingers neurotically on one of her still-fascinating thighs while we were talking to a particularly militant pro-tobacco senator and decided the latter was probably true. An opportunity to test that theory presented itself a few minutes later.

"Have you met Dr. Jacobs?" I asked, grabbing her by the arm and dragging her toward a man with gray hair and narrow, stooped shoulders. He was standing alone in the middle of the party and appeared to be having a quiet conversation with himself.

"Dr. Jacobs? How are you doing, sir? I'd like you to meet Anne."

He took a step back instead of offering his hand. It was nothing personal—he'd been doing that for as long as I'd known him.

"Carl Jacobs?" Anne said, her voice straining a bit as she tried to keep herself in check.

"That's right," I said. "Dr. Jacobs here is the head of Terracorp's Science Department. He's been studying tobacco for . . . well, for longer than you and I have been alive."

Truth be told, I was being a real asshole here. Carl Jacobs was a supernice guy who was terrified of nearly everything—particularly people.

"So, Doctor," Anne said. "What is it exactly you study?"

"Tobacco," he said nervously.

"Specifically, what's something you might be involved in?"

"Flavor."

"Flavor, really?" I could hear it in her voice—she was going to crack in about thirty seconds. It turned out that I didn't have to wait that long.

"Strange. I could swear that you're the man who discovered that putting ammonia in cigarettes gives smokers a bigger nicotine jolt and addicts them faster."

Jacobs took another step backward, trying to hide behind the drink in his hand. "Not me! I mean, I don't think there is any conclusive evidence of that."

"You know," Anne continued, "I read when they recalled Perrier because of poisonous levels of benzene, the actual amount in each bottle was one two-thousandth of what you'd get from a single cigarette. Is that true?"

I put my hand on Anne's back and she jerked her head around toward me, assuming I was coming to Jacobs's rescue.

"I still want to find out where those shrimp have gotten off to. Would you guys excuse me?"

I avoided Jacobs's eye and wandered away, making a mental note to wallow in guilt about it later.

After getting a club soda and a few more snacks, I avoided wading back into the crowd, instead skirting the edges of it and finding a shady spot beneath a tree. In the distance, I could see that Jacobs had managed to light a cigarette but that it didn't seem to be deterring Anne any. Finally, he made a break for it, cutting a swath through the well-dressed crowd while she relentlessly pursued.

The club soda was cold, and the shade and temporary solitude felt good. I'd grown up with these people, I worked with them, I had the same history as them, but I didn't feel any connection. It's strange not to have anywhere you belong.

I walked along a truly beautiful hibiscus hedge, finally stopping to watch a hummingbird dipping its beak into a particularly vibrant blossom.

"Pretty, aren't they?"

I spun around, but no one was there. I walked a little farther, and a bench that had been hidden by the hedge began to appear. Sitting on it was Paul Trainer.

"They are," I said, feeling the groundless calm that stemmed from being here with Anne suddenly abandon me.

He tilted his waxy-looking face toward the sun and closed his eyes. As near as I could tell, he was the only person at the party wearing a suit— dark gray despite the heat, with a jacket so long that you'd have to go back to the eighteen hundreds to find a time when it was in fashion.

"You were right, you know," he said.

"Excuse me?"

"About what your generation feels when they smoke. Information sets when you're young. When I was a kid, the danger was played down and cigarette advertising was everywhere. Even after everything that's happened, it's hard for me to believe they're bad for me . . ."

I nodded, but decided that the less I said the better off I'd be.

"When *you* close your eyes all you see is the warning labels and the public-health studies. You were too young to remember when the Marlboro Man could move and ride and smoke. In a way, I feel sorry for your generation. Information sometimes comes at the price of romance."

"People are more health conscious now," I said, already forgetting to keep my mouth shut.

He laughed. "Your generation worships at the altar of safety, doesn't it? 'Have a safe holiday.' 'Drive safe.' No one just wants to have fun anymore."

"I guess not."

The hummingbird was gone now and with it my reason for standing there, so I turned to leave.

"Why'd you pick history, Trevor?"

I stopped, not sure how to take the fact that Paul Trainer had bothered to find out my college major.

"The study of the past seems so . . . dead to me," he said. "Like studying shadows. This happened, that happened . . ."

The truth wasn't very impressive. When I'd arrived at college, I'd been typically unsure of what to do and I'd had a really good history teacher my first semester. It turned out to be the right decision, though. I really enjoyed it.

"It's not what happened so much as who it happened to, Mr. Trainer. It's the study of human nature. It's finding what was behind those dates and events that's interesting."

"What's behind the tobacco industry's dates and events?"

I shrugged. "America was founded on tobacco exports. World wars were fought on adrenaline and nicotine. The development of our political and economic systems was molded by tobacco money . . ."

"Did you know that the military used to put cigarettes in C rations?" he said, eyes still closed, face still thrust toward the sun.

"I didn't."

He smiled. "Yes you did."

In my peripheral vision, I saw Anne bearing down on Congressman Sweeny.

"Historical characters tend to become caricatures over time, don't you think, Trevor? Either very good or very evil."

I nodded, suspecting that he was spying on me from behind those old eyelids.

"You went to Duke, right?"

"Only for a little while, Mr. Trainer. I did part of a master's there."

"But you didn't finish."

"No."

"Why not?"

"I got kicked out."

"And how did you manage that?"

"The real reason or what they said?"

"The real reason. I know what they said."

"My master's thesis . . . well, it involved doing DNA tests on African-Americans who could trace their ancestry back to slaves working for wealthy southern families. I was interested in finding out how much Caucasian blood was mixed in."

He laughed—the first time I'd ever seen him do it. Honestly, it was kind of a frightening little ritual: His chapped lips curled back, revealing large, tobacco-stained teeth, then a loud, monosyllabic bark.

"What did they say when they found out what you were planning?"

"They said they didn't feel it, uh, meaningfully added to the information base."

"I would have given anything to read it," he said through the violent coughing fit his laugh had triggered. "But it couldn't possibly have been as brilliant as your memo on hiring Pamela Anderson as our primary spokesman."

"You . . . you read that?"

"Read it? Hell, it's still hanging above my desk at home."

I wasn't sure how to respond to that.

"And what about us, Trevor? What's our place in history?"

I wasn't sure how to respond to that, either.

"Something wrong, son?"

"I'm not sure I'm qualified to lecture you about the tobacco industry."

"This is a party, for Christ's sake! We're having a pleasant conversation. People do that at parties."

"I'm not sure I have anything to say that would be worth your time."

"Look out there," he said, pointing to the people covering my dad's lawn. "Exactly three-quarters of those people want to kiss my ass, and the other quarter wants me to kiss theirs. I'm not in the mood for either today. Anything I do that doesn't involve *them* is worth my time."

I figured I'd more or less made my bed at this point—and anyone I hadn't pissed off yet, Anne was probably taking care of.

"I think the tobacco industry is a victim of its history, Mr. Trainer. For a hundred years we've terrorized the federal and state governments— paralyzed them with fear. And for a long time that worked for us. But now with the rise of litigation . . . Well, that paralysis is coming back to bite us."

"You make it sound so . . . nefarious."

"Smooth denials aside, we kill nearly half a million people a year."

It was a brave statement, and I was glad I'd made it. It went a little ways toward reversing the Video Game Debacle, as it would henceforth be known.

"That's not entirely true, Trevor. We provide a product that allegedly allows nearly half a million people a year to kill themselves."

"Don't say that too loud or people will start going after us under the assisted-suicide laws."

"So, in your opinion, Congress and the president are just going to sit back and watch the courts tear us apart."

I shrugged. "The world has changed, and we haven't changed with it. We've gotten into the business of putting out fires . . ."

It suddenly occurred to me that I was maybe going too far. Trainer himself had directed that losing strategy for the last twenty years. I was trying to figure out a way to backpedal when Anne miraculously appeared.

"Trevor . . ."

"Anne, let me introduce you to Paul Trainer."

He jumped to his feet and took her hand, bowing a little as he shook it. "It's very nice to meet you, miss."

"You, too," she said, but something was obviously wrong. She seemed barely interested.

"Could you excuse us for a moment, Mr. Trainer?"

"Of course."

She pulled me along the hibiscus until we were out of earshot.

"Trevor . . . there's a problem with your mother."

"Really?"

She gave a furtive little motion with her head and I looked out over the crowd, instantly spotting the problem she was talking about. Mom's face was blurred with the hair that had fallen across it and her joints seemed to have loosened. I turned back to Anne, not needing to watch any longer. I'd already seen it.

"Maybe we should take her inside," Anne suggested.

I'd been trying to do that since I was thirteen. She'd hug me and make a loud speech about how I was always looking out for her, while everyone watched with those superior little smiles. Then she'd ask me to get her another drink.

"She'll be fine."

Anne seemed a little angry at my apathy and opened her mouth to yell at me but then didn't.

I heard footsteps on the grass behind me and turned to see my father and Paul Trainer approaching.

"I'm real sorry to interrupt," Trainer said to Anne. "But we've got a Montana update coming in over the videophone, and I was wondering if I could steal Trevor for a little while."

She offered a strained but polite smile.

My father, who clearly wasn't very happy, tapped his watch. "Now, Trevor."

Trainer put a hand on his shoulder. "We can wait a few minutes, can't we, Edwin? Say, five minutes, Trevor?"

I nodded dumbly as Trainer took my father by the arm and led him away.

"I'm getting the impression that you haven't been entirely truthful about your role at Terra," Anne said.

I only half heard her as I concentrated on the crowd parting to let my father and Trainer through.

"Trevor?"

"I honestly don't understand what's going on, Anne. I'd swear that I just spent the last half hour insulting the man . . ."

"I got everyone else."

I felt a smile spread across my face. "Thanks. And thanks for coming—it really helped." I handed her my keys. "You'll be happy to know that your work here is done. Get Jimmy to pull my car up for you. I don't know when I'm going to be able to get out of here."

"Do I *have* to leave?"

"No. Leave whenever you want."

"Then I'm going to stay a little longer."

And wait for me, I wondered?

"There are still a few people I want to talk to."

She started back toward the party without so much as a backward glance.

15 MONDAY MORNING WHEN I WALKED THROUGH THE office, everything went silent. People disappeared behind the chest-high walls of their cubes, ducked through doors, and hustled off toward the copy room. You'd think I would have found all this desperate scurrying interesting but honestly, it just made me tired. It was becoming harder and harder to maintain the illusion that I'd ever had the trust and friendship of the people I worked with and that they didn't just see me as a mildly dangerous curiosity. A dancing bear.

The meeting I'd been called into during my father's party had provided me with yet another hour of intense discomfort. Daniel Alexander had come off a little less smug and bored than when I'd met with him, and that sarcastic sense of humor my father had gone on about was nowhere in evidence.

The meeting did accomplish one thing, though: It gave some credence to Trainer's rather implausible story as to why he would send someone like me to check up on our Montana legal team. Alexander's report seemed to be less focused on the trial than it was on the reasons why his five-hundred-dollar-an-hour involvement was critical. I guess he was worried that someone in management might figure out that they could hire a forty-dollar-an-hour ambulance chaser to lose for us.

I'd spent the entire meeting sitting quietly in a corner, trying to ignore the looks of the other men in the room as they tried to figure out what I was doing there. Fading into the background had always been one of my greatest talents, but now I seemed to be failing even at that.

I sympathized with those men, though, because I wanted to know what I was doing there, too. Was it Trainer's goal to humiliate me—to subject me to people who reminded me that I was a know-nothing nobody? If so, it was working. On the other hand, maybe this was his idea of a reward for entertaining him in the garden.

Anne was gone by the time I emerged from the house, so I took a position by the shrimp bowl and tried to watch Trainer without being too obvious. I couldn't quite figure him out. Was he more cheerful since our doom-and-gloom meeting or was it just that the party had become more subdued around him?

When the cab Jimmy had rounded up for me dropped me at home, I'd found my car in the driveway and my keys in the post slot. There had been no note from Anne thanking me for a lovely time, no message on my machine from her implying that I wasn't such a bad guy after all, and no invitation to a quiet little dinner Sunday night.

There was, however, a photo of a naked girl diving into a star-shaped pool tacked to my door with an arrow. The note on the back, this time in the shaky scrawl of the man himself, said simply: "Silly boy." I hoped this little communiqué had appeared *after* Anne had been there, but the way my luck was running, it seemed kind of unlikely.

"Anything going on?" I asked Ms. Davenport who, to her credit, managed not to flee at the sight of me.

"This just came in." She handed me a printed e-mail, and I read through it twice. "Is this a joke?"

"No. The documents you're supposed to deliver are on your desk. Paul Trainer's assistant brought them down personally."

I AIMED MY CAR at an empty parking space and jumped out, hurrying across asphalt to a small, white office building. Because of all the dopey secrecy surrounding my involvement in Smokeless Youth, I'd never been to their headquarters and was forced to hunt around a little before I found a door stenciled with the letters *SY.*

The front room had the feel of a dentist's office waiting room, but without the five-year-old copies of *Woodworking* and *Black Entrepreneur.* A woman sitting behind an open window frame eyed me suspiciously, undoubtedly certain that I was some tobacco company mercenary sent to shoot up the place.

"Can I help you?"

"Yeah, I'm looking for Anne Kimball. Is she in?"

"Can I tell her who's here?"

"Trevor Barnett."

The surprise on her face suggested that she recognized my name, and I decided to use that to my advantage. A quick, surprise assault was probably best, as it wouldn't give Anne a chance to jump out a window.

"Don't bother," I said in as friendly a tone as I could muster and headed for a door that I assumed led into the bowels of America's teen antismoking effort. "I'll just head on back."

I glanced into each office as I moved along the poster-lined hallway, finally finding Anne sitting at her desk in the last one. I entered quietly and rose up on my tiptoes to get a better angle on what she was scowling at. It turned out to be a concept for SY's ill-advised "five million kids smoke" campaign.

"Hi, Anne."

Her head rose slowly, as though she was hoping the voice had just been a figment of her imagination.

"Trevor. What are you doing here?"

"I was wondering if I could borrow you for a little while?"

"Borrow me? What do you mean? What for?"

"I've got an errand to do for Paul Trainer, and I think you might find it interesting."

"Uh, thanks for the offer, but I'm swamped. We've got this thing—"

"I talked to John on the way over. He agreed that you should take a break and come with me."

She seemed a little angry—obviously not happy about me and John O'Byrne planning her day behind her back.

"Look, Trevor . . ." She pointed to the door behind me, and I closed it.

"I don't want to give you the impression that I dislike you. That's not the case. I mean, what's to dislike?"

It occurred to me that Jimmy had said almost exactly the same thing to me yesterday, and for some reason that bothered me.

"But, I don't think we're really on the same wavelength. I mean, I appreciate you taking me to the party and all—it was really educational. But . . ."

"This is pure business, Anne. I swear. And I think you'll enjoy it. In fact, I guarantee it. Come on, it's one day. I may not be the greatest company in the world, but if you grit your teeth, I know you can stand one lousy day."

ONCE AGAIN, I found myself in ambiguous territory. Was this assignment a reward or a setup? It seemed unlikely that Anne's rampage through the party would have escaped Paul Trainer's notice, and he would be rightfully angry. It seemed to me that I should have been fired by now or, at the very least, damned to a job well beneath upper-management's radar. Obviously, though, ulterior motives and hidden agendas abounded and I was caught somewhere in between them.

Anne was sitting directly across from me, arms crossed tightly in front of her chest and lips irretrievably sealed. Yesterday, she'd been fairly successful at feigning discomfort in the deep leather seats of my Lincoln Navigator, but she was having a hard time maintaining that pretense as she languished in the overstuffed loungers of Paul Trainer's jet. Oh, she tried: She fidgeted back and forth, peered out the window, flipped impatiently through magazines. But there was no escaping the fact that this was the only way to travel.

"Where are we going?" she said finally.

"I can't tell you. It's a surprise."

Long silence.

"How's your mom?"

"I'm sure she's fine. Thanks for asking."

Another long silence that I couldn't figure out how to fill.

Finally, she waved a hand around the jet. "You keep telling us you're just a paper pusher. Is this a family perk?"

"I thought that after the party you'd realize there are no family perks."

"A lot of people have sticky relationships with their parents, Trevor. But not everyone inherits a gazillion dollars."

I laughed. "If I tell you something, will you promise to keep it to yourself?"

She thought about it long enough that I figured I could trust her. "Okay."

"My grandfather put a bunch of tobacco stock in a trust for me when I was born. As long as I continue to work for Terra, I get a fairly small annual distribution based on dividends and capital gains."

She thought about that for a moment. "But there aren't any capital gains or dividends. There haven't been since . . ."

"Not since I gave SY that big check," I said, completing her sentence. "The silver spoon in my mouth isn't as big and shiny as everyone thinks it is."

"Why do you stay, then? Why not go get a better job?"

A fair question.

"Well, the other provision of my trust is that it gets distributed to me when I'm sixty."

"That's a long time away," she said, pointing to the pack of cigarettes in my pocket.

I shrugged. "I know it's probably hard for you to understand, but this has been my life for as long as I can remember: Go to work for the company, get payments, be rich at sixty. It's not that easy to walk away from. It should be, but it's not."

She looked a little skeptical. "Are a few payments that you're not even getting anymore and the remote possibility that you might get some cash before you die worth all this moral discomfort? I've got to wonder."

"What discomfort?" I said a little too quickly.

"Get real, Trevor. If you're so happy with your lot in life, why are you involved with SY?"

"Whether or not I shuffle papers and collect my money doesn't make any difference as to who starts smoking and who dies," I said, clumsily avoiding the question.

"Fine. Whatever."

Long silence number three.

"I once tried to get transferred to Terra's food subsidiary," I admitted finally.

"What happened?"

"The trust administrators wouldn't go for it. Besides, I'm not all that sure it would have solved my problem. When we got turned down, my lawyer sent me an article from *Newsweek* about everybody in America killing themselves with junk food. I must have read it a hundred times. Turns out that if you work at it hard enough, you can snap your neck with a feather pillow."

"How can you even compare cigarettes and junk food, Trevor? Tobacco provides no benefit to anyone, and the industry has been knowingly killing people for years for the sake of nothing more than money. It's a sinister industry. Downright Orwellian."

"I don't know," I said, too shocked by the fact that we were having a conversation to be properly afraid of where it might lead. "If anything's Orwellian about it, it's people's ability to doublethink. I mean, despite all the information out there about how dangerous smoking is, people still manage to close their eyes and do it."

"Because your company spends billions to obscure that information."

"Come on, Anne. That dog just don't bite anymore. There have been warnings on the packs for longer than we've been alive. You can try all you want, but you're not going to convince me that people don't know smoking is bad for them."

She shook her head in a combination of wonder and disbelief. Then she started clapping. "You may be the most brilliant rationalizer I've ever met."

"I don't think that's fair. I—"

"Of course it's fair, Trevor! People lie to themselves—they always have. It's human nature. 'If I buy this shampoo or this Bowflex, I'll look just like the model. If I drive this car, or wear this cologne, women will fall all over me ...'" She pointed at me. "'I'm just a lowly file clerk, I'm not responsible.' You take advantage and you know it."

There's nothing like arguing with someone who's right to make you say stupid things you don't mean.

"I guess there's a sucker born every minute."

Anne sank a little farther into her seat. "Try all you want, Trevor, but you don't play the hard-ass convincingly. You like P. T. Barnum? Me, too. Do you know what else he said?"

I just sat there.

"He said, 'Money is a wonderful servant but a terrible master.'"

Not a moment too soon, the copilot appeared in the cockpit door and started walking toward us with a cell phone held out in front of him.

"Mr. Barnett? I'm afraid we're being diverted to L.A."

"What? What do you mean we're being div—"

He proffered the phone and I snatched it from his hand. "Hello?"

"Trevor! Sorry to spring this on you, boy." Paul Trainer's voice.

"I don't understand. We're not going to see ..." I glanced up at Anne and cut myself off.

"Damn, son ... I know how much you kids like those rock and rollers, but we got an emergency on the West Coast and I need you to handle it for me."

"An emergency? What kind of emergency?"

"What? I can hardly hear a thing you're saying. These goddamn cell phones! I swear the world would be better off if we all used a couple of Dixie cups and a real long string."

"I *said,* what kind of—"

The line went dead, and I reluctantly handed the phone back.

"Things aren't going as planned?" Anne said as the copilot wandered back to the cockpit.

"No."

I flopped back in my seat and began chewing my thumbnail.

"So we're not going where you thought we were going?"

"Uh uh," I said dejectedly.

"Then there's no harm in telling me where that was. Right?"

I sighed quietly, digging a sealed manila envelope out of my briefcase and holding it up. "Paul Trainer asked me to deliver some papers to Ian Kingwell. I know he's your favorite singer, and I thought you might think it was fun to meet him."

I didn't really expect a display reminiscent of the Beatles' first landing on American soil, but I did expect *some* reaction to my statement. Instead she just sat there with her brow slightly crinkled. Finally, "What kind of papers?"

"I dunno. I guess Terra's food division is sponsoring his concert series. Something about that."

"The food division," she said. "Didn't we just establish that you don't work for the food division?"

I shrugged, and she snatched the envelope from my hand.

"Hey!" I said as she leaped from her seat and ran to the rear of the plane, tearing at the envelope as she went.

"Stop that!" I yelled, chasing her. "You can't open those. They're confidential!"

She crammed herself into a corner with her back to me, and I tried unsuccessfully to reach around and retrieve the pages that she was now skimming.

"Blah, blah, blah," she said, and then tossed a page casually onto the floor. I fell to my knees and scooped it up.

"Anne!"

"Blah, blah, blah . . ." Another page jettisoned. Then "Wait . . . Now, you might find this interesting. It says here that in addition to prominently displaying company logos, Ian is supposed to use company products of his choice on television and in photographs." She turned and looked down at me. "What products do you figure those would be, Trevor? You think maybe he'll pause during an interview for a vitamin-rich helping of instant Mac and Cheese?"

"I don't know!" I protested.

But of course I did. While I'd been careful not to look at the documents or give their contents any conscious thought, Terra's goal here wasn't all that complicated. Ian Kingwell was a balls-to-the-wall chain-smoker who for some reason almost never indulged his habit publicly. The death of Kurt Cobain (suicide) and the rise of the oh-so-nonsmoking Britney Spears crowd had left the industry in desperate need of good, old-fashioned, rock-and-roll puffers.

"You should be ashamed of yourself," Anne said, flicking the rest of the pages at me in such a way that they scattered all over the floor. "That's not just a bunch of ink, Trevor. Those papers are going to kill people. And I'll bet you know just how many, don't you? I'll bet your accountants came up with a number so they'd know how much to spend."

"We don't—"

"You don't?" she said, cutting me off. "Well then why don't you and I try to come up with a number on our own? Doesn't that sound like fun?"

"Not rea—"

"Let's see. You figure twenty million kids see Ian with a cigarette on TV. We'll be really conservative and say one in ten thousand decides to take up smoking because of it. That's what? Two thousand kids? I'll be generous and say that half quit and only twenty percent die from the habit. So . . ." She tapped her front tooth with the nail of her index finger, and I couldn't help wondering if that's where the chip came from.

"Congratulations. That's two hundred people. Not a bad day's work . . ."

"Jesus Christ, Anne! I didn't negotiate the deal!" I shouted. "His manager and Terra hashed this thing out months ago. And now I'm not even going to deliver the documents! I have nothing to do with this."

"Of course, I'm being insensitive. It's not your fault. You never actually do anything, do you? Other than stand around and pretend you don't exist."

I rose to my feet, and she jabbed me in the chest. "Do you know that when I met you, I thought there was something in there? I mean, sure, you were completely lost but at least you seemed like you were *trying* to find your way. I figured that was more than can be said for most people. You really had me fooled."

16 "LOOK," I SAID, MAKING A SHOW OF LEANING OVER the seats and glaring at the side of our driver's face. "Seriously. We want to know where you're taking us."

He returned a vapid smile but kept his eyes dutifully on the road. "I'm sure they'll explain everything when we get there."

"Who's 'they' and where's 'there'?" Anne demanded.

"I really can't say, ma'am."

"You can't or you won't?"

"Anne, this really isn't that big a—" I made the mistake of putting a hand on her shoulder, and she picked it up and tossed it away as if it were a dead rat.

Her frustration wasn't hard to understand. I'd had a lifetime to grow accustomed to the tobacco industry's obsession with secrecy: "I have no idea what you're talking about/You can't prove a thing/There are conflicting studies..." Ambiguity and feigned ignorance were our sword and shield. Her problem was that she was trying to process what was happening using logic. In my experience, this approach was almost always doomed to failure.

She peered through the window as the driver swung the car into a well-kept industrial park and then skidded to a halt in front of a windowless building with FOX NEWS stenciled on the door.

It seemed like I was supposed to get out, so I did. Anne followed, albeit reluctantly.

"Mr. Barnett!"

A woman wearing an absolutely Puritan business suit, complete with tall white collar and a brooch burst from the building, grabbed my arm, and began physically pulling me along behind her. I glanced back and confirmed that Anne was keeping up.

"It's wonderful to meet you. Really, a privilege. I'm sorry that we don't have time for makeup—"

"Makeup?"

"Don't worry! You look great! We were hoping you'd have a tailwind, but sometimes things don't work out, you know? But it's fantastic that you're here. Fantastic. Having a completely unanswered interview like this would have been horrible. A disaster. Particularly now. But then, I don't suppose I have to tell *you* that."

I managed to get a hand out and keep a set of double doors from swinging back and hitting me in the face as I was dragged through them. The section of the building that we entered was cavernous and unfinished-looking—temporary walls, cables, and metal spotlights anchored to a concrete floor. I'd never been in a TV studio before and while it was kind of exciting, I couldn't shake the feeling that this was a bad thing. I finally stopped short, causing the Puritan lady's arm to slip from mine as her considerable momentum carried her forward. Anne ran into the back of me.

"Why am I here?" I said. "And who *are* you?"

She gave an exasperated shake of her head. "You haven't been briefed? I'm so sorry. I'm Cynthia Bates—I work in Terra's L.A. public-relations office. We got a last-minute call telling us that Fox was interviewing Angus Scalia—you know how the press like to nail us at the very last possible moment—and they asked us if we wanted to provide an industry spokesman to be on the show with him." She gave my arm another tug, but I didn't move. "And none of our local people were available. Good thing you were close by."

"I wasn't close by."

She gave me one of those "You're telling me more than I want to know" shrugs and moved behind me to see if she could break my inertia by pushing instead of pulling.

"Uh, I think there's been a mistake, here. Cynthia, is it? You see, I'm a file clerk, not an industry spokesman. I've haven't been on TV since my Little League team won the state championships."

"You're Trevor Barnett, aren't you?"

"Yes, but—"

"It's my understanding that Paul Trainer called *personally* to notify our office that you were being sent here to give the industry's side of the story."

"I don't really know what the industry's side of the story is. I think maybe Mr. Trainer made a mistake."

She actually gasped at my blasphemy, which would undoubtedly be reported in triplicate before the day was over.

"The show's airing in just a few minutes, Mr. Barnett. We have to hurry."

"I don't think you understand. I don't want to be on TV. You go. You're in public relations. Isn't this the kind of thing you do for a living?"

Cynthia blinked a few times before she spoke, enunciating clearly and slowly. "Mr. Trainer requested that you do this *personally*."

"No."

I felt Anne's hand on my shoulder, and she leaned around me with a less than reassuring smile on her face. "What Mr. Barnett means to say is that he's very excited about the prospect of appearing before millions of viewers and explaining his position on the tobacco industry." She added her weight to Cynthia's, and I found myself being propelled inevitably forward.

"Okay, Mr. Barnett," a girl who looked fresh out of college said as she shoved me backward into a chair and crammed an earpiece into my appropriate orifice. "The format here is pretty informal. Mr. Flag"— the host, who I'd watched tear apart everyone from Enron executives to environmentalists trying to save baby seals—"will ask both you and Mr. Scalia questions, and you should answer them as clearly and concisely as you can. Try to avoid talking over each other—no one will be able to understand you."

I looked around, but all I could see was lights and cameras. "Where is he?"

"Who? Mr. Scalia? He's on a remote feed from Miami. You won't be able to see him, but you'll be able to hear him."

"Can I get some powder out here?" she shouted and then leaned over me to make sure I was connected to everything. I breathed in, hoping to find the comforting tobacco scent of a potential ally but could only smell shampoo.

There was a brief moment of idleness, and what was about to happen here suddenly came fully to roost.

"You know, I don't think this is such a good idea," I said, trying to stand.

"You're hooked up!" the girl protested and shoved me back again. A moment later another woman appeared and without warning smacked me in the face with a big powder puff. I squinted at her through the sudden fog as she examined her handiwork. "That should keep some of the shine off. Try to relax, okay?" Then she disappeared past the lights.

Flag rushed in a moment later, ignoring the flurry of activity around him as he took a seat behind his desk. We shook hands, and he had the good manners to try not to be obvious when he wiped my sweat off on his slacks.

"Okay, we're coming in from commercial," a disembodied voice said. "Three, two . . ."

At first, I didn't have to say much: "I appreciate you having me," was my relatively smooth answer to Flag thanking me for being there. Scalia was a little more opportunistic, launching into an angry diatribe almost immediately.

Honestly, I couldn't tell you a thing he said. Being on TV is a lot harder than people make it look. All I could think to do was stare wide-eyed into the camera, or at Flag, or at my feet. It wasn't lost on me that everything I said and did was probably going to make me out to be either stupid or guilty to the millions of people watching. Obviously, it wasn't lost on Anne, either. I could see her at the edge of the set, arms crossed and feet spread wide. She seemed to be enjoying watching me forced out from under my rock and into the sunlight.

"The bottom line here," Scalia said as I tried to focus and quit thinking about Anne, "is that we are finally witnesses to the long-overdue destruction of the tobacco industry. The Montana suit is an enormous step forward, but it's only the first of many. The cigarette pushers will finally have to pay for their crimes. Their assets will be seized and distributed to the families of the people they've murdered—"

Flag cut in, looking directly at me in a way that made the sweat running down my spine go cold. "Is that your take on the situation, Mr. Barnett?"

I've never watched the tape of this interview, so I can't say how long I sat there, silently staring past Flag at Anne. If I had to guess, I'd say a good ten seconds—which is the equivalent of about fifty hours on TV. As that time ticked by, Anne seemed to become less and less firmly planted to the floor. Then she started to look a little concerned. Finally, she made a motion with her hands, trying to prompt me to say something.

But what? The mealymouthed platitudes and legally defensible denials that Paul Trainer and she expected me to spout? For the first time in my life, I couldn't rely on my uncanny talent for fading into the background. I actually had to do something.

"Mr. Barnett?"

"No," I heard myself say. "That's not the way I see it."

"Of course not," Scalia replied. "They never do, do they? They see everything through a filter of smoke, blood, and money."

I wasn't exactly sure what that meant, but the hatred in his voice was dead honest. And the only way I saw to combat that kind of sincerity was with an equal amount of it.

"If you ask me to predict the future, I'd have to say that we would bankrupt ourselves and the plaintiffs' attorneys would settle for a much lesser amount in order to get paid. Then, we'd raise prices on cigarettes and the smokers would cover the cost of the judgment."

"Mr. Scalia," Flag said, nodding. "I have to agree with Mr. Barnett here. His prediction seems fairly realistic to me."

My confidence swelled a bit at having survived the first exchange.

"Of course. But what Mr. Barnett isn't talking about are the four similar suits in process. And the twenty more that'll be filed when we win in Montana. And make no mistake—we *will* win. What are you going to do about those, Mr. Barnett?"

I shrugged. "It's hard to say. We're already paying over two hundred billion dollars to the states—more than our total profits from nineteen fifty to the present. I suppose that in order to keep handing out money to smokers, we'll just have to keep increasing prices on cigarettes. It'll be kind of like Social Security—current smokers will pay the tab for past smokers."

Scalia didn't seem to have a set answer to that and moved off point a bit to give himself time to think. "Cigarettes have been a scourge on society for hundreds of years. It's been proven that the industry has lied repeatedly about the dangers of smoking and that the government has colluded with them. Where the government is concerned, the average citizen lost his voice years ago. The same people are elected over and over and are slaves to corporate special interests. But the public can still speak through the court system. What Mr. Barnett won't tell you is that with every judgment being followed by an increase in price, cigarettes will eventually become so expensive that no one will want to buy them anymore. And then the empire will crumble."

"It's a great conspiracy theory," I said. "But it seems to me that the government's attitude toward smoking is very much a reflection of the public's attitude on the subject."

"I WOULD EXPECT Mr. Barnett to react that way," Scalia said. "He comes from a long line of tobacco people going back almost to the beginning of our country. His family is responsible for more deaths than the Third Reich."

Flag tried to hide a little smile. References to Hitler always made good television. I, on the other hand, felt angry. Furious, really. Why? I'd been called a Nazi before.

The answer came to me in another long pause that I've decided to label as dramatic and not stupid. The truth is, I wasn't a Nazi. I wasn't even good enough to be a Nazi. I was just one of those flaccid dumb-asses who moved into the Jews' houses and ate off their dishes, and read the books on their shelves; all the while being very careful not to consider where all those Jews had gotten themselves off to.

They were being killed? they would say, stabbing at a piece of strudel with a silver fork etched with some dead man's initials. *My goodness? How was I to know?*

Now you would think this would be a really disorienting realization—a defining moment, as they say. You'd think that I would jump up and proclaim the evils of tobacco or repent my sins against the hapless creatures who shared this beautiful planet with me. Unfortunately, I didn't really feel any of that. All I felt was an overwhelming sense of disgust that went beyond just myself and grew to encompass Scalia, the media, Paul Trainer, my father. Pretty much everyone living or dead.

"It seems a little hypocritical for Mr. Scalia to call me a Nazi when he's the one sitting here trying to dictate his values to the rest of America—to take away their freedom," I said, strangely energized by my nonepiphany.

"Their freedom? Do you mean their freedom to let you murder them? Or are you more concerned with your own freedom to make a product that kills hundreds of thousands of people every year and costs this country billions in medical expenses?"

"You know as well as I do, Mr. Scalia, that the cost to the government is just an accounting fantasy. The antitobacco lobby loves to tell anybody who'll listen that smokers die ten years younger than nonsmokers, but no one wants to give us credit for doing our patriotic duty and expiring before the most unproductive and medically expensive years of our lives.

The truth is this: Tobacco is a huge financial boon to this country. But maybe you're right. Maybe we should ban cigarettes—take them away from the informed adults who choose to use them. I have to wonder if they should be the first things on the chopping block, though. I mean, they're not even the most dangerous product we make."

Flag leaned forward across the desk, obviously interested in this new twist on the normal smoking debate. "And what *is* the most dangerous product you make, Mr. Barnett?"

"I would have to say those boxes of little doughnuts—you know, the ones with the powdered sugar on them?"

"This is ridiculous!" Scalia shouted. "The fact that people like you can make light of the horrible deaths of hundreds of thousands of people is . . . is . . . sinister."

"I'm not making light of anything, Mr. Scalia. Are you aware that diabetes in people in their thirties almost doubled over the last decade, as has the incidence of obesity? Even using the inflated death numbers published by the antitobacco lobby, obesity has taken over as the number-one cause of preventable death in America. And just as bad are the ten to twelve thousand dollars a year in treatment that diabetics need—something you don't have with smokers. Of course, that doesn't include the cost of complications like amputations, blindness, heart disease, and absenteeism. The total cost of diabetes to America is almost a hundred billion dollars a year—the same as all cancers combined. And that number is nothing. It's going to explode in the next few years as the children who are now being victimized grow up."

"I refuse to debate—" Scalia started. But I was on a roll as I'd never been before. I know it sounds strange, but Scalia had come to personify everyone who had ever put me down and every insult I'd ever swallowed. I cut him off again, surprising myself by looking the show's host directly in the eye.

"Did you know that smokers are less likely to be overweight than non-smokers? No one ever mentions the catastrophic health problems that smokers avoid by being thin."

I realized after I spoke that I really didn't know if it was true, but it sounded good and it's not like people hold the tobacco industry to a particularly high standard of integrity.

"Mr. Flag, are we going to waste your valuable time while Mr. Barnett tries to divert attention from his own horrible crimes?"

"And why single out doughnuts?" I said, ignoring him. "Why not also get rid of fatty foods and candy, too? I think it's been thoroughly proven they're far more deadly and costly than cigarettes. And while we're at it, I think there would be a lot less disease and reliance on the health-care system if everyone had to exercise three days per week. We could make it a law. Every couple of months, everyone in America would go down to their local town hall and weigh in." I pictured the mound of soft flesh at the other end of my earpiece. "If they're overweight, we could just fine them or make them pay their own medical costs. There. I've just saved the government a trillion dollars."

"You're very smooth, Mr. Barnett. Very clever. But have you ever seen a person die of lung cancer? Watched them wither away to nothing, choking on their own blood? That's what you're selling. Not only death. But a horrible death."

I didn't argue. What was the point? It was all true.

"Mr. Barnett is obviously terrified," Scalia continued. "He inherited a personal fortune that my sources say is one hundred percent invested in tobacco stocks. He lives off the capital gains and dividends paid to him every year. If there are none, then he has to work like the rest of us."

I heard the words, but didn't immediately process them.

"Mr. Barnett?" Flag said. "Is that true?"

How had Scalia gotten that information? I immediately thought of Anne but then realized she'd been with me ever since we'd talked.

My rage and the temporary illusion of invulnerability faded and I became reacquainted with the millions of people watching me.

"I did inherit tobacco stock," I stammered. "But it's not much money."

I heard Scalia laugh. "Not much money to a tobacco industry executive. But I think that the average American who's out there working his tail off to put food in his kid's mouth would consider seven million dollars to be a fair amount of cash."

My momentum had completely disappeared, and I wasn't smart enough yet to just shut up.

"It's nowhere near that much now! And there are no dividends or capital gains, so—"

"But if you can shirk your responsibility to your victims and increase sales there would be, wouldn't there? You'd be rich again."

"I—"

"Just knowing that there are people like you in the world, Mr. Barnett, terrifies me. Knowing that there are people who would involve themselves in the deaths of millions of people simply to try to protect some investment income. Let me say to everyone out there watching this television show: You're looking at the personification of the tobacco industry. Stop smoking and you may survive. Keep smoking and Mr. Barnett here gets a yacht."

"That's not true!" I said, way too loudly into my microphone.

"Really, Mr. Barnett. Tell me why it's not true."

How could I? How could I explain my entire life, my inheritance, my father, my job? How could I explain things that even I myself still hadn't managed to reconcile in my mind?

"You know what?" I said, yanking my earpiece out and standing. "I don't need this crap."

"ARE YOU all right? You don't look so great."

Out of the corner of my eye, I could see Anne watching me with an expression that made me think of someone examining a mutilated animal by the side of the road.

"I'm fine," I said, pulling a cigarette from a pack on the dashboard and lighting it. I didn't want to ask her to leave because I had it in my mind that this would be the last time I'd ever see her. I figured I'd let the fumes do it for me.

"Why do you do that?"

"What?"

"Smoke."

"Because it's expected."

"Is that really it? Or is it because you believe that if you kill yourself at the same rate as everyone else, you're off the hook for all this?"

"Please, Anne—enough, okay? I just lost my job and my trust in the course of half an hour. Maybe you're right and in the grand scheme of things, the scales aren't balanced. But today I think I'm even."

"You'll find another job. It probably won't be as hard as you think."

"Are you kidding? I just got on national television and proved beyond a shadow of a doubt that my phone number is on Satan's speed dial. I'm guessing that the offers aren't exactly going to be pouring in."

She opened the car door and stepped out, letting in a gush of clean, damp air. As I watched her walk around through the headlights I realized that even after everything that had happened, the foremost thing on my mind was Ian Kingwell.

On our almost completely silent flight home, I'd started thinking about my real motivation for taking Anne to meet Kingwell. What had I hoped to gain? Sure, I'd told myself that it would make her like me more. Was that the truth or had it just been part of a subconscious plan to demonstrate that I wasn't so bad. To show her that even the great Ian Kingwell was for sale?

What was I turning into? I had my faults, but I'd never been a mean person. But now, after a few days of cavorting with upper management, I'm siccing Anne on the defenseless Dr. Jacobs and then stripping her of her heroes.

I rolled down the window. "Anne?"

She stopped and turned back toward me.

"I . . . I'm sorry about today."

I couldn't read the expression on her face, but I could see that she was concentrating on me harder than she ever had before.

"Were you listening to yourself today, Trevor? When you were on with Scalia?"

"I don't know. Not really, I guess. I was pretty nervous."

"That's a shame. Because some of what you said was worth listening to."

17 I<small>T WAS MY FAVORITE CARDBOARD BOX: COMFORTABLE</small> handles, spacious, sturdy. I'd emptied it of Christmas ornaments last night and was now carrying it like a shield as I weaved through the cubicles toward my office. The timidity that had recently overtaken my colleagues seemed to have dissipated a bit, and they uniformly lifted their heads and watched me pass.

I'd hit my all-time low the night before and honestly thought I might be on my way back up at that point. For sure, I was scared but I wasn't really depressed. The people I worked with had already proven that they'd never thought of me as anything more than an oddity, and I'd already mourned the loss of them. Anne seemed kind of perplexed where I was concerned, which I took as a substantial improvement over the disdain she'd so energetically displayed before.

Today, as they said, was the first day of the rest of my life.

I was within sight of my office door when I suddenly found my narrow path blocked.

"Hey, Stan. What's going on?"

"What the hell were you thinking?" For the first time in a week he was actually expressing an honest emotion to me. My performance on TV and the box in my hand must have finally cleared up any confusion about my status at the company.

"I—"

"What the hell were you thinking?" This time he almost shouted the words.

There was a muffled rustling behind me as people came to the doors of their cubes to watch.

"We've got the courts breathing down our neck with a judgment that could put us out of business and you make jokes about us killing our customers and then suggest outlawing cigarettes?"

"Come on, Stan. You—"

"But, hey what do you care if we go down the toilet? You've got your seven million dollars, right? But I don't!" He indicated around the office. "None of us do! What we've got is kids to put through school and mortgages—things you wouldn't know anything about."

I stood my ground but found myself taking an increasingly defensive posture behind my box. The irrational panic of a person facing the end of his familiar routine could be powerful. I knew that as well as anyone.

"What I said on television isn't going to make any difference to the jury in Montana or what happens to your position here, Stan."

"You don't believe Scalia and the others can do anything to us? I guess you can afford to believe whatever you want."

"Maybe not as well as you think," I said, and started around him.

I was surprised when he stuck out a thick arm and blocked me.

"Tell you one thing about Scalia, though. He calls 'em like he sees 'em. You've been running around this office for years trying to prove you're one of us, and in ten minutes he showed you for the spoiled child you are."

I looked behind me at the faces of the people watching us. They weren't the faces of my best friends or soul mates but they were the faces of people who'd sat in my office and BSed after hours, people whose homes I'd been to, people whose kids' names and accomplishments I knew.

Stan withdrew his arm and I continued to my office, feeling a little disoriented. Everything that was familiar to me was crumbling, and I wasn't sure if there was going to be anything behind it.

"Good morning, Trevor."

Ms. Davenport had her own box—not quite as nice as mine, but already filled with the contents of her desk. I realized that I hadn't ever considered how this would affect her.

"Ms. Davenport, I—"

As was often the case, she didn't let me finish.

"Sue Jensen in Accounting offered me a job, and I took it."

"She's a nice lady," I said, genuinely happy that Ms. Davenport had found a way to land on her feet. I watched her pick up her box and walk away. And that was it —the end of a two-year relationship. Strangely, I've never laid eyes on that woman again.

IT TURNED OUT that I didn't need my box.

When I walked into my office, it was pretty much empty. My chair was still there, as was my desk, but with all the drawers open and cleaned out. My mug, my national park calendar (this month had been Yellowstone), my CDs, my day planner—all gone. Even my posters espousing the health benefits of lighting up as often as humanly possible and the patriotism of Luckys had disappeared, leaving only slightly grayed outlines on the wall.

I sat down, trying to fight off a growing sense of embarrassment. Everyone out there had watched Security pack up my stuff, and they were all just sitting there waiting for me to be escorted out. At this point, it would probably be smarter to just slink away under my own power.

I was almost out the door when my phone, now residing on the floor, started to ring. I hesitated for a moment and then picked it up.

"Hello?"

"Trevor?"

"Anne?"

"I was wondering what happened. I tried your house, but there was no answer. I wasn't sure if you'd be going into work or not."

"Me neither."

"I guess everyone saw you on TV."

I looked around my empty office. "I guess so. I'm just waiting for Security to come and throw me out."

There was a short silence over the line. Finally, "How does it feel?"

At first I thought she was being sarcastic, but her tone sounded genuine. I turned my back to the door and lowered my voice.

"I honestly don't know. A little numb, I guess. The status quo hasn't ever been great for me, but it was familiar, you know?"

"What are you going to do now?"

"Find a job, I guess. You guys wouldn't be hiring by any chance, would you?"

She laughed at that. "I don't know if I'd come around here right now if I were you. John's watched the video of you and Scalia about ten times. I think he's still trying to figure out whose side you're on."

"He ought to be happy. Scalia made me and everybody else in the industry out to be complete assholes."

"I don't know, Trevor. I think John's favorite part is where you implied that by being fat, Scalia was part of the problem. He'd never admit it, though."

"What do you think—"

I hadn't heard anyone come up behind me, but the phone was suddenly snatched from my hand and slammed down into its cradle.

I spun in my chair, ready to defend myself against a gang of security guards but found myself faced with much worse.

"Dad . . ."

"What the hell would possess you to go on national television and suggest that damn near everything Terra makes be outlawed?"

"I—"

"And then just walk off and leave Scalia to sit there for another fifteen minutes and drive his point home unchallenged?"

I had to scoot my chair back so that I could look up at him without hurting my neck. For some reason it didn't occur to me to stand.

"The stuff about my trust—"

"You want to play with the big kids, Trevor? The big kids go out and find information like that. They bribe people, they dig through garbage cans . . ."

That was the problem. I didn't want to play with the big kids.

"Do you know how many phone calls I've gotten this morning?" my father continued. "I've talked to the governor, three senators, and I don't even know how many attorneys. The head of our food division called me at home and he's already projecting a two percent drop in their snack cake division's sales next month."

In retrospect, the only thing that kept me from giggling at a grown man worrying about people eating fewer Ho Ho's was that the man was my father. No matter how badly your relationship deteriorates and how estranged you become, it's hard not to think of your father, on some sub-conscious level, as Santa Claus, Albert Einstein, and Alexander the Great all wrapped up in one.

The phone began ringing again, but I didn't dare answer. Anne probably thought I'd hung up on her. Great.

"I know, Dad. I know I did a lousy job, okay? And I shouldn't have walked off like that, but people aren't going to give up their Ding Dongs because of what I say any more than they're going to give up their cigarettes."

"So I'm overreacting?" Suddenly my father's stylish tie looked like it was strangling him. He lifted the *New York Times* that for some reason I hadn't noticed in his hand and began reading a front-page piece.

"Tobacco Industry Spokesman Suggests Outlawing Cigarettes."

"That's not fair. It's taken out of conte—"

"Fifty years of legal strategy," he shouted, waving the paper around in the air. "And you decide to just undo it in a day."

"Come on, Dad, now you *are* overreacting. I didn't undo anything. Any politician who even jokes about outlawing cigarettes would get crucified."

My father seemed as surprised by the irritation in my voice as I was. He gestured around my empty office. "Well, look where your little tantrum's gotten you. I suppose it's my fault for giving you too much. You never had to grow up; you never had to do anything for yourself. Maybe if you'd have been forced to struggle for something, you'd have more self-respect."

"Maybe," I said, a little dismissively. I just didn't need this right now.

His eyes narrowed. "Did you ever stop to think how this makes me look? Let me tell you. Between you and your mother I look like an ass."

And with that observation, he spun around and disappeared through my door. So much for the possibility of reconciliation with my father.

It was clearly time for me to run away before Security showed up, but instead I just sat there thinking about all the years I'd spent at Terra and how Paul Trainer had finally gotten his revenge on me for defying his Godlike authority. He hadn't just fired me; he'd made me out to be an evil, spoiled bastard in front of the whole world. I looked around me again. And where the hell was my stuff? A lot of it wasn't company property—it was mine. What gave them the right to take it?

I slid my hand across the top of the desk where I'd kept my mug. It was shaped like a dinosaur's head and had been given to me by my mother when I was only a kid. The history of my life was dyed into it: the red of childhood Kool-Aid faded to pink from my binge-drinking years in college and now slowly being covered by the brown of coffee. Someday, I wanted it to be stained with pureed vegetables or whatever it was you ate when you were a hundred. It was mine. And I wanted it back.

That stupid mug, in a matter of seconds, became a symbol as powerful as the American flag to me. I jumped out of my chair and nearly ran out of my office, heading straight for the elevator. About halfway there, someone said something that I didn't catch but that was obviously directed at me. I stopped short and spun around.

"What? What was that?"

Silence.

"WHO SAID THAT?" I shouted.

No one answered.

"Yeah, that's what I thought."

I SHOT OUT of the elevator and started banging my hand on the thick wall of bulletproof glass that Terra's management worked behind. The woman on the other side seemed startled initially, but then I saw her reach beneath her desk for something. The alarm button, I assumed.

Then there was an unexpected buzz and click as the door in front of me unlocked.

"Mr. Barnett—" the woman said, but I ignored her and ran past.

"Hello, Trevor. You—" Trainer's secretary said, but I blew right by her too, forcing my way through the CEO's door.

Strangely, he was standing in the middle of his enormous office, hands on hips, as though he was in deep thought. He looked up when his door bounced off the wall.

"Where's my mug!"

"What?"

"It looks like a dinosaur. It's mine and I want it back."

He spread his wrinkly palms out in a gesture of peace. "Um. Okay."

"You had no right to throw me to the dogs like that."

"You had the jet and you were close," Trainer said, moving over to his desk and taking a seat on its edge. "They gave us three goddamn hours to get someone out there or they were going to put Scalia on alone."

"And now you're making it my fault."

Now all this might seem inordinately stupid, but this whole sensation of being really pissed off was kind of appealing to me and besides, what more could Trainer do to me? I mean, how much more screwed could I possibly be?

"Well, I have to admit that you didn't really stick to the script, but you showed a hell of a lot of passion."

I must have looked like I was ready to shoot up the place, because he suddenly took on the soothing tone you'd use on a child in the grips of a tantrum. "Now, take it easy, Trevor. You seem upset right now. Why don't we go get you that mug of yours and you can have a nice cold drink out of it. How would that be? Would that make you feel better?"

He slid off his desk and put a hand in the small of my back, pushing me gently toward a side door to his office, where *surely* there was a Security team with a butterfly net in just my size. Reality turned out to be even more strange.

The office we walked into was a mirror image of Trainer's and was decorated with essentially the same deep wood paneling, understated art, and plush sofas. It had been vacant ever since the death of his executive vice president and childhood friend (emphysema, though everyone said stress).

"There it is," Trainer said, hurrying me over to a large desk covered with my stuff and handing it to me.

"That's a good-looking mug, son. I can see how you'd want to keep ahold of it."

He wandered over to a cozy conversation pit and motioned to the chair in front of him.

"Want something in that? Some juice? Water?"

I shook my head and sat down.

"You're sure?" he said, putting his feet on a slate-topped table that had black marks exactly matching the heels of his shoes.

"Yeah."

"You know, Trevor, I've been feeling like things have been getting away from me lately. Either I'm not as quick as I once was, or there's just a whole lot more to think about . . ."

He checked to see if I was listening, and I nodded to prove I was.

"I was thinking I could convince you to help me out. But you're right— I shouldn't have had your things brought up. I shouldn't have presumed."

I just sat there.

"So what do you say, Trevor? Are you interested in the job?"

A little more sitting.

"Trevor?"

"I don't think I understand what you're asking me, Mr. Trainer."

"Call me Paul."

I felt my eyebrows rise. You could count the number of people who called him Paul on one hand. There was wide speculation that his own mother had referred to him as "sir."

"I don't think I understand what you're asking me, Paul."

"And I'm not sure how I can be more clear. I want you to work directly for me. To help me out."

"Uh . . ."

He bared his teeth in a wide smile. "Goddamn obesity statistics. I almost wet myself. Have you seen the footage? That fat son of a bitch looked like you'd hit him with a bat. I'm pretty sure I heard America let out a loud cheer. Where'd you come up with that diabetes stuff? Is it true?"

"Most of it. I think."

"Doesn't matter. That's the great thing about statistics—you can make them say anything you want. If the press calls you on anything, let me know and we'll get our guys to cook something up."

"I heard the CEO of our food division is kind of upset," I said. "I guess he's worried about their cupcake sales."

"Jesus Christ! I've got a multibillion-dollar industry crumbling around me, and that guy sits around worrying about a two percent drop in the sales of Twinkies. If he calls you direct, you transfer him straight to me. Moron . . ."

"You know, Paul . . ."

A voice floated in from his office. "Mr. Trainer?"

"I'm in here."

Richard Horton, the company's CFO, appeared in the doorway. I couldn't help noticing that he called Paul "Mr. Trainer."

"How are we doing, Rich?"

Horton was the opposite of what you'd expect from a guy who could keep the numbers of a huge multinational corporation in his head. He was good-looking, relaxed, and well liked by just about everyone who'd ever met him. I'd always found him a little intimidating.

"Right now we've got a one and a half percent drop in stock prices more or less across the board," he said, standing in the middle of the room with his hands shoved in his pockets. "It'll probably get a little worse, but not much."

Then another interesting thing happened. My father came in through the same door. He moved slowly up behind Horton, staring at me as I clutched my mug.

"What's Wall Street saying?" Trainer asked.

"Trevor's comments worried them," he said matter-of-factly. "Not so much the talk of delegalization—more the loss of control on national television. Add that to the Montana suit and it didn't exactly stop the erosion of confidence in the industry."

I was trying to calculate what a one and a half percent drop in stock prices meant in real money, but stopped counting before there were enough zeros to make my mouth go dry.

My father didn't. "So we're talking about hundreds of millions of dollars here?"

Trainer winked at me. "I guess we ought to try and keep you off the TV for a while, eh Trevor?" He turned back to Horton. "How are the diabetes-related stocks doing?"

"Up. You ought to do pretty well."

"Thanks, Rich."

Horton left, but my father stayed.

"Have a seat, Edwin," Trainer said, pointing to a chair. "Do you mind if we use your office for this, Trevor? I'm having a new computer installed in mine." He nudged me with his foot. "I'm on level four." No doubt a reference to Darius's new game.

To his credit, my father managed to take the subtly delivered news that his screwup son had an office adjoining Paul Trainer's without any overt breast-beating or hair rending.

"Edwin, last week you told me there was an eighty percent chance that we were going to lose the Montana suit. I want to know what we're going to do about that. Trevor here says we bankrupt ourselves and settle with those bastard lawyers for a few billion."

"You understand, Paul, that Trevor has no legal training . . ."

"I don't give a shit about his legal training. Do you agree with him or not?"

My father managed a smile that didn't look too strained. "The plaintiffs' attorneys are going to want to be paid. They'll take a deal."

"And on that day, we throw open the floodgates. . . ."

My father looked like he was about to say something, but Trainer held his hand up.

"And that's just the class-action suits. What about the individual suits? How many are we fighting right now?"

"Somewhere around twenty-five hundred."

"Jesus." He faced me again. "What's your crystal ball say, Trevor. What's our future?"

Out of the corner of my eye, I could see my father glaring at me like he had so many times throughout my childhood.

"We drown, I guess. The law and public opinion—our culture—are moving against us. They have been for a long time."

"You're not a lawyer, Trevor," my dad said again, as though that was some basic flaw in my character.

"That wasn't a legal opinion."

"He's right," Trainer said. "If you take enough wild shots you eventually hit something. Our stock price is going to continue to slide, we're going to continue to have to write checks to these whiny assholes. Pretty soon we're not going to be able to raise capital or get loans." He shook his head slowly. "We've got the hearing on the new surgeon general's report day after tomorrow, and we don't have a lot of friends on that panel. They smell blood and votes, and they're going to be grandstanding. Insufferable bastards."

He leaned back a little farther into the soft tapestry of the sofa and suddenly seemed to be talking to himself. "We can't just sit back and fight a defensive war anymore. No, not anymore. It's time to take a stand."

18 THERE WERE ABOUT TWENTY PEOPLE MOVING ALONG the buffet, trays gliding in front of them on a set of chrome rails. I watched Anne from a distance as she balanced a carton of milk on top of the mountain of food she'd created and shot the people in front of her annoyed looks as they agonized, read labels, and asked poignant nutritional questions.

Having already eaten—in the executive dining room—I took a seat at an empty table and continued to spy on her as she happily grazed her way toward the cash register. According to a mathematical theory called the Heisenberg Uncertainty Principle, it was impossible to observe something without affecting it. My relationship with Anne was absolute proof of Dr. Heisenberg's genius. When I was around, she always seemed a little bit on guard—though perhaps not so much against me as against herself. She seemed horrified by the possibility that she might let a smile appear on her face or that she might show me something beneath that all-business façade of hers.

She paid and then scanned the cafeteria for an empty table. When her gaze crossed my part of the room, I waved. Just a casual little wave—nothing that could be prosecuted under the stalker laws. More of a "what a coincidence that I happened to be in this part of town, miles from my office, in this particular restaurant, just when you were having a late lunch," kind of wave.

She squinted at me for a moment and then walked over with perhaps a little less hesitancy than I was used to.

"You hung up on me."

"Actually, I didn't. My father snuck up behind me and grabbed the phone. Sorry."

She looked around but as luck would have it, there were no more tables. I motioned to the chair in front of me, and she took it.

"You're still wearing a suit," she observed, dumping her silverware out of its bag and attacking something that might have been Asian noodles.

"There's been a slight hiccup in my plan."

"You had a plan?"

"Maybe *plan* is too strong a word. But after my performance yesterday, I was pretty confident that I was going to get fired."

"I take it you didn't," she said through a full mouth.

"Huh uh. They just moved me out of my old job to a new one."

"What are you doing now?"

"I'm Beelzebub."

"Excuse me?"

"Uh, I'm working directly for Paul Trainer. He gave me the old executive vice president's office."

She didn't seem surprised by any of this. To her, it was all just part of a conspiracy I'd been involved in from the start.

"Congratulations."

"I didn't want the job."

"But you took it, didn't you?"

"I guess."

"So now you're the assistant to the CEO of Terracorp," she said coldly. "Wow. That's an important job. I'm sure you'll do well."

This wasn't working out the way I'd fantasized. The glimmer I'd seen in her yesterday, and that I was here hoping to fan, was almost too faint to see now.

"It's all kind of an accident," I said, sounding a little desperate. "I mean, I took you to that party and let you chase people around and insult them,

I gave the board a report that consisted of ten words. Last night I got on TV and completely threw the party line out the window, then walked off the stage. Trainer must be getting senile . . ."

"I think you're selling yourself short, Trevor," she said in a tone suggesting she wasn't even close to buying what I was telling her. "I would think that a guy who doesn't believe in anything and spends all his time trying to convince himself that he has no responsibility for anything would be perfect for that job. Maybe Paul Trainer knows *exactly* what he's doing."

"I . . . I don't think that's fair."

"What part?"

Good question.

"Have you had a chance to think about what you said on TV yesterday?" she asked, pushing her tray away from her for the moment. "Have you thought about where you want all this to go? Or are you going to just drift along with your eyes closed?"

I didn't answer.

"If this sounds mean, I'm sorry, I don't mean it to, but you're like a sail: empty unless you're filled with whatever wind happens to be blowing at the time."

I would say I felt deflated, but that would sort of support her sail analogy, wouldn't it?

"In a way, I kind of envy the opportunity you've got, Trevor. At the very least, you have ringside seats. Won't it be interesting if the industry loses and those seven hundred thousand Montanans are mad enough that they don't let their lawyers talk them into settling? Won't it be interesting if they end up taking everything the industry has?"

I was feeling a little of that unfamiliar anger again when I finally spoke up. "Maybe more interesting than you think."

Her expression turned a bit guarded. "Why do you say that?"

I stood, a little stiffly. It was way past time to walk away.

"What do you figure they'd do with all those assets, Anne? Dismantle them and sell them for ten cents on the dollar? Or would they make a deal with Paul Trainer to keep running things and fill their bank accounts with the profits? Moral outrage can sometimes get lost in these kinds of numbers."

She surprised me by grabbing my arm as I began what I thought was as dignified an exit as could be hoped for under the circumstances.

"Trevor, can you see that you may actually be in a position to *do* something here? To maybe influence the way things will be in the future—"

"I'm not in a position to do anything," I said. "Trainer decided for some reason to make a show of asking my opinion and pretending to listen, but I'm not stupid enough to believe it."

"Are you sure you're not just telling yourself what's easy to hear? No one expects anything from a bystander, right? But what if you could save one life. Just one? That'd be pretty heroic."

I pulled my arm free. "It's more complicated than you want to admit, Anne."

"Maybe."

I looked down at her, and she didn't turn away. "Why don't you have dinner with me, Anne? Give me a couple of hours to change your mind about me. I can't guarantee the quality of the company, but I can promise you some really exceptional food." I pointed at her tray, still piled high. "And lots of it."

That made her smile, but she tried to hide it with her napkin.

"You know, Trevor, I don't doubt that somewhere under all that confusion, you're probably an okay guy. And, if it doesn't make me sound bad saying it, you're really beautiful. But I don't think so."

19 THE NEXT MORNING, I FOUND MYSELF SITTING BE-
neath a large portrait of my grandfather, at a gigantic table
occupied by the CEOs of America's major tobacco companies. With the
exception of Trainer, they had a fairly uniform look to them: sixtyish; con-
servatively dressed in gray suits and white or blue shirts; short, dark hair.
All spoke with upper-crust southern accents that tended to degenerate a
little when they got angry—though only about three-quarters had the
earthy hoarseness that used to be a prerequisite for the job.

Trainer was the only man standing. He was wearing one of those
hundred-years-out-of-style suits he favored in a color that, depending on
the light, might have been purple. At least ten years older than anyone else
in the room, he paced persistently through the haze generated by the cigarette in his hand.

"I'm not sure all of you know Trevor Barnett."

I nodded, forcing myself to make eye contact with everyone in the
room and trying not to dwell on how out of place I felt.

"I've created a new position—executive vice president of strategy—
and Trevor's agreed to take it on. I've asked him to sit in with us."

It was the first time there had been any mention of a title for me and
while I thought it sounded pretty impressive, no one else seemed to. It was
likely that their impressions of me had already been formed based on my

brilliant work on television. Not a single one of them acknowledged me in any way. Trainer paced for a few more seconds and then stopped short.

"We're going to lose this suit and we're going to have to settle."

He seemed to be daring the men in the room to disagree with him, and no one took him up on it.

"We don't have any friends out there anymore. The politicians are frozen, the antitobacco forces are on the attack, the nonsmokers are bitching about secondhand smoke, and our customers think we're over here twisting our mustaches trying figure out ways to kill them faster." He went into motion again. "The world's changed—legally, politically, economically—and we're still playing the same game we were twenty-five years ago. Have I missed anything, Trevor?"

I glanced up from the notes I'd been taking and saw that Trainer was looking directly at me.

"I'm sorry?"

"Son, we've got people to do the minutes. Have I missed anything?"

Everyone turned toward me.

My initial reaction was typical: Who was I to talk at this meeting? Then I came to the odd realization that the combination of my job going through industry documents, my genealogy, and my degree probably made me more knowledgeable about the history and philosophy behind the tobacco industry than anyone in this room. Maybe I did have something worthwhile to say.

"I think there's a question we need to ask ourselves: *Why* don't we have any friends left?"

"You tell me," Trainer said, making enough of a show of listening to me that the others didn't dare show their irritation.

I cleared my throat. "I have to wonder if a big part of it is that we spend so much time trying not to make enemies that we don't have any time left to try to make friends. People get behind offense, not defense. How can anyone rally behind an industry that's still half denying cigarettes are bad for you and winning lawsuits by hiding behind technicalities? I mean, we've always done well for ourselves, but it's generally been through manipulation and legal tricks. No one respects either of those things. People get excited about winners," I said, moving into territory that I knew too well. "Not people who spend all their time trying not to lose."

"I'll buy that," Trainer said, with what could have passed for a respectful nod. "So what do we do, Trev?"

That was a lot harder question to answer. "I'm not sure there's anything we can do. We've bought politicians, scientists, media—you name it. We've ruthlessly bullied anyone who's gotten in our way. In the end, though, we did too good a job. Like you've said before, the politicians are terrified. Of us. Of smokers. Of the antismoking lobby. Of tobacco farmers. And after a hundred years of cultivating that terror, it's backfiring on us." I needed to take a breath but was afraid of losing my momentum, so I plunged forward. "The industry didn't anticipate the judicial branch running the country."

"It's a disservice to this great country," Trainer said. "If America was founded on anything, it was personal responsibility, self-reliance, and tobacco."

"That's true. But it was also founded on capitalism run amok: the trusts, the robber barons, slavery, child labor. We aren't going back to those times. In fact, the current is pretty strong in the opposite direction."

Chuck Fay, the head of RJR and a tough old bastard who'd done multiple tours of duty in Vietnam because he was "having such a damn good time," spoke up.

"But what the hell does that mean, exactly? I mean, I'm on board with you and I'm goddamn well ready to get off my knees and kick some ass. But how?"

Honestly, I didn't know. I wasn't sure how best to say it, but the point I'd been trying to make was that we were probably doomed and should all be looking for jobs.

"And whose ass?" another man said. I wasn't sure which company he ran. "Trevor here keeps telling us what's wrong and not offering a solution."

I felt my heart rate jump at the criticism, but Paul Trainer came to my rescue.

"It's not Trevor's job to fix this industry—he's just here to provide information. The buck stops with us."

Everyone settled back into their chairs, and I once again faded comfortably into the background.

"As I see it, there are two things we need to focus on. The first is that, while people are pretty cynical about the tobacco industry, they're just as cynical about politicians. Those bastards have been running around criticizing us all over the press and then getting into bed with us every night for years. That has to stop—we've got to drain the political currency out of that kind of shit."

"Just tell me how," Chuck Fay said, obviously excited about the prospect of a stand-up fight.

"Action, Chuck. It's the one thing that terrifies politicians. It's easy to spout off about morality as long as it doesn't affect anyone's life, but when it does, the people who were happy with the status quo get pissed." He gestured dramatically. "Everyone's turned against us because it's easy to turn against us. But when push comes to shove, I'm not all that sure the majority hates us all that passionately. Take nonsmokers. They hate us because they see us as the evil empire we've been portrayed to be and because they figure we cost taxpayers a bunch of money. But they aren't directly affected by smoking in and of itself, and frankly no one's stupid enough to believe they're going to get a tax cut if the tobacco industry dries up and blows away. What about smokers—our customers? They hate us because it's easier to blame us for killing them than it is to blame themselves. But in the end, they sure don't want some fat bastard like Angus Scalia taking their cigarettes away."

"You were talking about action," Fay said. He was the only person in the room who didn't seem intimidated by Trainer. "What action? How can we control twelve asshole Montanans sitting in a jury box?"

I didn't know Trainer well, but I'd have to say that he seemed pleased with the way the meeting was progressing. The smell of panic was in the air. Stock options, retirement benefits, prestige—everything these men held dear—was starting to realistically look like it could just go away one day.

"Make no mistake, gentlemen. We are at war," Trainer said. "And we're losing. The enemy has been chipping away at us, and now we find ourselves weakened and outnumbered."

The way he was speaking and his emphatic gesturing reminded me of the beginning of the movie *Patton*. But instead of a huge flag as his background, I imagined a more potent symbol of our country: a giant dollar bill.

"As I see it, we have one last chance at a counteroffensive. Every day we delay reduces our chances of victory."

Honestly, this was starting to get a little silly. But there was something about Trainer's schizophrenic charisma that made it almost work. He paced back and forth a few more times and then came around the table, approaching me with his hand outstretched. I shook it, and he slapped me on the back and leaned into my ear. "You did a hell of a job, son," he said quietly. "I'm proud of you. Now if you could excuse us . . ."

I have to admit to feeling a little pride as I walked from the boardroom. For the first time since all this began, I hadn't come off as a complete idiot. That small step in the right direction seemed to be masking the vague queasiness I'd felt since my last (in more ways than one, perhaps) conversation with Anne. So I focused on it.

TWO HOURS LATER, the board was still locked in its meeting. What were they talking about? Why had I been asked to leave? I gave those questions about five minutes of thought before putting my mind to a more pressing question. What was it exactly that the executive vice president of strategy did?

I wandered around my office running my fingers over the rich surfaces, paced it off (thirty by forty feet), put a few things away, and then finally just sat down behind my expansive desk. Unfortunately, my computer wasn't on line yet, so no surfing and I'd given Trainer my only copy of Darius's new game. A few more minutes and I found myself staring out my door at the empty desk just outside.

That was it! What new executive vice presidents did was hire assistants. But how did one go about that? Pretty much everyone I knew at the company had made it clear they hated me—though in light of my new title, I was guessing they were having a change of heart on that point.

Honestly, I needed more than an assistant. I needed an ally. I was used to my moral compass being pretty stable (okay, stuck), and right now it was spinning a little bit out of control.

I now see what I did next for the stupid and desperate act it was, but if there is one great truth that history has taught us, it is that many great things have been accomplished with stupid and desperate acts.

I dialed the phone and put on the wireless headset, then stood and began walking around the edges of my office.

"Smokeless Youth."

"Anne Kimball, please."

"Can I ask who's calling?"

"Trevor Barnett."

I was put on hold for a moment, and then the woman came back on. "I'm afraid she's not here. Can I take a message?"

"Why don't you transfer me to John O'Byrne."

"One moment."

I examined a large plant by the door while I waited. I couldn't for the life of me figure out if it was fake.

"Trevor! How are you? I saw you on TV—quite a performance. Anne tells me you're reporting directly to Trainer now? That's exciting news. Very exciting. How did it happen?"

"I'll let you know as soon as I figure it out myself. Is Anne there? I need to talk to her for a sec."

"I think she's in her office. Hang on and let me check."

I waited another thirty seconds before Anne's tired and irritated voice came on the line.

"This is Anne."

"Hey, it's Trevor."

"John told me."

She was obviously angry with me for not taking the hint that she didn't want to talk.

"I'm not sure how else to say this, Trevor, but—"

"Before you say anything, let me tell you that this isn't a social call."

"It's not?"

"Remember what you said about being jealous of my ringside seats? It looks like I've got an extra ticket."

"I don't understand."

"I'm offering you a job. I need an assistant. You'd sit pretty much right outside Paul Trainer's office and watch all the important comings and goings. You'd be able to sink your fangs directly into the industry's jugular . . ."

Silence.

"Anne? Are you still there?"

"I'm here."

"Well? What do you think?"

It took a few moments for her to answer. "I'm not sure what you're asking me, Trevor. What's the goal here?"

"What do you want it to be?"

"I'm not sure there's anything I could accomplish there, Trevor. I mean, we know everything we need to know about the industry. My challenge is getting people to do something productive with that information . . ."

"What about helping me? You could do that."

She exhaled loudly in what might have been a laugh. "Help you do what?"

"We'd have to work that out."

"I . . . I don't think so, Trevor. I just don't see myself sitting behind a desk at Terra."

"What about all that stuff you said about how I might be in a position to affect the way things are? I think you're being a little selfish, don't you?"

me save just one person's life?" I
:he day before. "Wouldn't that be
ır part?"

ʀoughout
Just deter-
al, and his
s look like
other side
to be a lit-
y to a con-
presence,
ıdly, care-
watching

ıder that
ɔnship to
elieve af-
ever felt.
trol and

s atten-
ıgton's
ch.
1 to go

20

PAUL TRAINER HADN'T SAID A WORD TH[
the drive, but I wouldn't say he seemed nervous[
mined. He'd opted for a more conventional suit than norn[
arms were folded across his chest in a way that made his elbov[
they were going to poke through at any moment. Sitting on th[
of the limo's wide rear seat was my father. Now, he did appear[
tle worked up. Whether it was the fact that we were on our wa[
gressional hearing about the new surgeon general's report or my[
though, I couldn't be sure. When he talked to me it was in a frier[
fully easygoing tone that sounded so wrong that I found myself[
his lips to see if they matched up with the soundtrack.

Quite a change from last time we'd met and yet another remi[
everyone—even my own flesh and blood—gauged their relatic[
me based solely on my status at the company. I know it's hard to b[
ter the week I'd had, but I think this was as uncomfortable as I'd[
I actually caught myself fantasizing that our driver would lose con[
run us into the Potomac so I could swim the hell out of there.

As we got closer to our destination, my father began to focus h[
tion on Paul. I squinted out the window at the bright white of Wash[
monuments as they flashed by, trying to remember which was wh[

"We don't have much more time," I heard him say. "We nee[
over a few things."

"Selfish?"

"Definitely. What if you could help me save just one person's life?" I said, paraphrasing her words to me the day before. "Wouldn't that be worth a little moral discomfort on your part?"

The line went dead.

20 PAUL TRAINER HADN'T SAID A WORD THROUGHOUT the drive, but I wouldn't say he seemed nervous. Just determined. He'd opted for a more conventional suit than normal, and his arms were folded across his chest in a way that made his elbows look like they were going to poke through at any moment. Sitting on the other side of the limo's wide rear seat was my father. Now, he did appear to be a little worked up. Whether it was the fact that we were on our way to a congressional hearing about the new surgeon general's report or my presence, though, I couldn't be sure. When he talked to me it was in a friendly, carefully easygoing tone that sounded so wrong that I found myself watching his lips to see if they matched up with the soundtrack.

Quite a change from last time we'd met and yet another reminder that everyone—even my own flesh and blood—gauged their relationship to me based solely on my status at the company. I know it's hard to believe after the week I'd had, but I think this was as uncomfortable as I'd ever felt. I actually caught myself fantasizing that our driver would lose control and run us into the Potomac so I could swim the hell out of there.

As we got closer to our destination, my father began to focus his attention on Paul. I squinted out the window at the bright white of Washington's monuments as they flashed by, trying to remember which was which.

"We don't have much more time," I heard him say. "We need to go over a few things."

"Same old crap," Trainer observed and joined me in enjoying the view. "It's not just—"

Trainer waved a hand regally and we rode the rest of the way in silence.

"You've got to be kidding me . . ."

Trainer banged on the glass separating us from the driver and it slid down. "We're not getting out here. Take us around to the back entrance."

"There is no back entrance, sir," the man said. "I was told to drop you here. Someone will meet you just inside those double doors over there and take you to the hearing room."

Trainer leaned over me and looked out the heavily tinted windows. There were about fifty people on the sidewalk milling around in a haphazard spiral, some carrying signs pointing out the industry's crimes against humanity, others pumping their fists in the air and shouting slogans. As usual, the people whose lives were enhanced by the mild, relaxing flavor of our product had opted not to show their support in person.

"Goddamn those prima donnas!" Trainer spat, undoubtedly referring to the congressmen we were going to meet. "I'll bet they found a back entrance!"

The crowd was starting to become interested in the limousine sitting silently alongside them, but hadn't yet broken formation to investigate.

"If we're going to go, we should go now," I said, shoving the door open, "while we've got surprise on our side."

I was recognized the moment I exited the limo, and the loud boos started immediately. Trainer slipped in behind, sandwiching himself between me and my father as we pressed through the crowd. I kept an eye on the tight corridor of people that formed for anyone who seemed to have been especially deranged by the death of a loved one or the cancer spreading unchecked through their internal organs. I found Anne instead.

She wasn't one of the sign toters and she wasn't booing. She just stood there, looking a little sad. I began to slow as I approached but felt Trainer's bony hand in my back. "Jesus Christ, boy! Move your ass! These people are out for blood!"

THE HEARING ROOM wasn't what I'd expected. No Romanesque monuments to Abe Lincoln or paintings of George Washington. No marble archways with Latin slogans across them. It was more like the Fox studio I'd been in or one of those old movie sets where everything was just a façade propped up by two-by-fours. Appropriate, I suppose.

It kind of felt like a wedding when we walked down the only strip of floor not filled with spectators. Trainer, feeling safer now, took the lead and walked with what seemed like complete confidence. All eyes were on him, and I was grateful to go more or less unnoticed.

Near the front, there was a half-full row of chairs that had been blocked with a red rope, and I recognized the other industry CEOs despite their uncharacteristically pale and shiny faces. They didn't seem to share Paul's sense of inner calm, though I honestly couldn't understand why. They were accustomed to being held up as the model of modern evil, and Trainer was the only one of them who had been called on to testify. Honestly, this surgeon general's report was just the latest in a long line of nonevents in the history of the industry.

I figured I was supposed to sit in the row with the CEOs since they were the only seats available, and I slowed to step over the rope.

"You're with me," Trainer said.

"Huh?"

"You're with me, Trevor. Edwin, why don't you hang back? I don't want to go up there surrounded by lawyers."

It's hard to describe my father's reaction to that. What is it they say? If looks could kill? But it wasn't directed at Trainer—he had his back turned. It was directed at me. I pulled my leg back over the rope and caught up with my new boss, taking a seat next to him at a table with a microphone on it.

I tried to mimick Trainer's slightly bored and authoritarian aura, though I doubt that it worked. Good practice, though, and it helped me forget the fact that my father was trying to glare a hole in the back of my head.

About five minutes passed before the congressmen began to file in. The chairman called the hearing to order with a few strained pleasantries and then dove right in with a minispeech about the wisdom of the American people and how organized, peaceful civil disobedience could bring about meaningful change. Honestly, I didn't see the connection, but the congressman's words did seem to affect Trainer. His posture straightened, his skin looked a little tighter, and his eyes shone a little brighter.

Finally, the unashamedly political soliloquy ended and the congressman turned his attention to me.

"Could you introduce yourself, sir? Are you Mr. Trainer's attorney?"

I tapped the microphone. "Uh, no sir. I'm—"

"This is Trevor Barnett," Trainer said, saving me yet again. "He's an executive vice president at Terra and my personal advisor."

According to a research piece I'd skimmed, Carl Godfrey, the chairman of this hearing, was a deeply religious man who'd been married forever and had three successful children. A generally honest and moderate Democrat who was clearly influenced by America's current dark mood regarding all things tobacco. Which I supposed he should be. By the people, for the people.

"Sir, I assume that you've read the new surgeon general's report," he said after the introductions were done.

"I've read a detailed summary by my Science Department," Trainer replied, neglecting to mention the legal report and my brilliant, if concise, historical perspective.

"And what did your Science people conclude?" Godfrey asked. The other politicians seemed content to just listen.

"Not much, really. They pointed out a few discrepancies in statistics and samples. Mostly they just dumbed it down. This kind of technical stuff tends to go right over my head."

That statement was regarded with suspicion and disdain by nearly everyone on the panel. They saw it as the beginning of the ignorance defense popularized and perfected right here in Washington. The one exception was Congressman Sweeny, who seemed to have recovered from his multiple run-ins with Anne Kimball at my father's party. He was nodding sagely.

"Well, I *did* read the report," Godfrey said. "I also looked through some of the prior reports and the tobacco industry's responses to them. I have to say that they made me . . . tired. As near as I can tell, there are thousands of studies that all seem to pretty much agree that smoking is absolutely terrible for you. That it causes cancer, emphysema, heart disease, and a hundred other things. Not to mention the addictiveness of it. And with every new report, these conclusions become more and more certain."

"The industry has conceded that cigarettes are dangerous and addictive," Trainer pointed out.

"That's true," Godfrey admitted. "But only recently. Why is that?"

"Sir," Trainer started, "there have been incredible scientific strides since this issue first came up. I'd point out that the early studies were not only inconclusive but often pointed to smoking being good for you. To this day, the process with which cigarette smoking supposedly does damage is not completely understood. The first surgeon general's report back in the sixties did move closer to putting the dangers into a more reliable context but the studies were certainly not perfect. We had some fairly zealous antismoking campaigners doing things like shaving rats' backs and painting them with nicotine. We even had one group give beagles tracheotomies and teach them to smoke. We've also had prior surgeon generals who've made absolutely false statements about the industry. In our minds—and obviously in the minds of the public who continue to use our products—the evidence was not conclusive."

He sounded great. Ultimately reasonable, calm, benevolent, and just a little bit frail—like a kindly grandfather. Unfortunately, not everyone was as impressed as I was. I tend to only recognize politicians who've been caught sleeping with their interns, so I didn't know the name of the man who suddenly cut in.

"I have to put a stop to this ridiculousness. All I want to know is how can you sleep at night knowing that you produce a product that kills hundreds of thousands of people every year?"

"I suppose the same way you sleep at night knowing that you've never so much as hinted at wanting to outlaw that product—because people should be free to make their own decisions about their health and because it's a cornerstone of the American economy."

Godfrey cut in before a shouting match could start. "I'm sure we all appreciate the history lesson, but I have to wonder where you're going with all this."

"If you could just indulge me for a few more moments, sir. . . . As I was saying, our scientists have gone over this report with a fine-toothed comb. And they tell me they believe that scientific and statistical methods have caught up with the problem."

I was with Godfrey—where was he going with this?

"So, in a nutshell, our people are in agreement with the surgeon general's conclusions."

The room suddenly grew loud as the spectators all began talking at once. I glanced back at the CEOs, and found that about half had turned from bedsheet pale to a kind of greenish gray. My father looked like he was going to vomit.

"Would you expand on that, sir?" Godfrey said, sounding as though he thought he hadn't heard correctly. Where were all the smooth denials and obscure legal arguments that defined these meetings? It no doubt sounded to him—like it did to me—that we'd just given away the farm.

"I'm not sure what more I can say, sir. We agree with the report." Trainer cut himself off there, but his tone and body language finished his thought for him: Now what are you going to do about it?

"WHY DIDN'T YOU tell me what you were going to do?" My father was talking to Trainer, but looking at me. I could see him in my peripheral vision as I watched the memorial to one of America's earliest tobacco farmers, Thomas Jefferson, speed by. The sun was setting on the city, giving it a kind of a magical quality that its inhabitants probably never noticed.

"We could have gone over your testimony, polished it. . . ."

Trainer frowned deeply. "What you're talking about is putting in a bunch of *allegedly*s and taking everything else out. That misses the whole point, doesn't it, Edwin? I didn't want this to seem lawyered to death. Everything that's come out of our mouths for the last twenty years has been sterilized by a pack of lawyers. People see right through that."

"But it protects us down the road," my father protested. A little stupidly, I thought.

"Protects us from what, Edwin? Lawsuits? I spent six hundred million dollars on lawyers last year and I'm not sure I wouldn't have been just as well off donating it all to a bunch of antitobacco zealots. I figure I'd be in the exact same position I'm in today."

"Are you sure you know what you're doing, here Paul?" my father said, sounding a little insulted. "Have you talked to the board about this?"

"Hell yes, I've talked to the board. I've talked to all the goddamn boards." In his reflection in the window, I saw him look at his watch. "You hungry, Trevor? We're out early."

"Out early" was something of an understatement. The hearing had disintegrated after Trainer agreed with everything in the report. A few congressmen tried to bait him with other angles, but he kept throwing it back in their faces by agreeing with their analyses and asking them what exactly they would propose to rectify the situation. Action. The one thing politicians feared as much as hidden video cameras.

"Sure," I said. "I wouldn't mind getting something to eat. Do you know a good place?"

"Paul," my father said, "there's going to be a huge backlash from this. We need to talk about how we're going to handle—"

Trainer grabbed a remote control and turned on the VCR in the console between us. "Have you seen this, Trev?"

He pushed Play, and an MTV interview with Ian Kingwell appeared on a tiny television screen. I leaned forward and watched him pontificate about the plight of children in the third world through the delicate spirals of smoke rising from the cigarette in his hand.

Despite a fair amount of effort on my part, I hadn't been able to forget Anne's overly kind estimate of how many people Kingwell's on-screen smoking would kill. The fact that I hadn't ended up delivering those documents seemed to have completely lost its relevance. I would have delivered them. I would have in a second. And that, if anything, was the bottom line.

21 I'D BEEN TO A TENT REVIVAL ONCE, BACK DURING MY brief and ultimately unsuccessful search for God. While it had been an energetic—perhaps even ecstatic—affair, I'd found myself a little put off by its lack of direction. Darius had shown his support by dropping some acid and joining me, but the chaos freaked him out and we left before he started speaking in tongues.

I know it sounds as if I didn't give God a fair shake, and that may be true, but it was starting to look like a viable substitute was about to be provided.

It all was pretty much like I remembered: the rolling farmland, the bright white of the tent canvas, the buffet of rich southern food (pies on a separate display, elevated so as to be closer to heaven). Behind the stage, Terracorp's logo had taken the place of the more customary worn wooden cross but was glowing in the afternoon sun with the same promise of bliss, serenity, and belonging.

We were about an hour and a half outside of D.C. in the surprisingly vast and rural landscape of Virginia, but I still didn't know why.

"What is all this, Paul?" I asked, slapping a mosquito against my neck and leaving a smear of my blood on my hand.

He pulled the tent flap back a bit farther, and I could see the men and women inside smoking free cigarettes and chatting impatiently.

"Remember back when you could count on the press?" Trainer whispered, ignoring my question. "No, I guess you wouldn't. You're too young. There was a time, though. A time when they didn't just go back and forth with public opinion—when they didn't kill stories that went against their customers' sensibilities or were too hard to get across in a ten second sound bite . . ."

"You mean back when we completely controlled them with advertising dollars instead of only half controlling them with advertising dollars?"

He looked back at me and smiled. "That's exactly what I mean."

The sun was beginning to set, and the tent's shadow now stretched out to infinity. I followed it with my eyes, trying to find my father, who seemed to have disappeared.

"Are you ready, Trev?"

"Ready for what?"

He thumbed toward the tent, a mischievous smile spreading across his face. "For them."

A funny thing about mischievous smiles. While they're cute on children and cool on college students, they tend to be a little sinister when worn by old men.

"You're not a spectator today, Trevor. This is your show."

Strangely, his statement didn't surprise me. Maybe I was starting to get a little jaded. The thought actually improved my mood slightly. I'd always aspired to be jaded.

"You're the one who made those statements to the committee, Paul. Don't you think you should be the one who does the Q and A?"

"This isn't a Q and A, son, it's a press conference. And CEOs don't do press conferences."

"Yes they do."

"Well, this one doesn't."

"Are you sure this is a good idea, Paul? Remember what happened last time I got on TV?"

"Oh, that was just a case of the jitters. This is going to be easy." He handed me a stack of three-by-five cards. "Just read 'em like they're written. And remember eye contact—with the cameras, too. It'll make you seem more honest. And if you feel yourself getting nervous, just picture the audience—"

"In their underwear," I said, dejectedly. "I know."

"I was going to say dead. Who would want to see these people in their underwear? That's just sick." He pulled the flap still wider. "I'll watch from here."

IT WAS HOT inside, and I started to sweat as I walked up an aisle lined with hostiles for the second time that day. The sun was beaming directly through the tent's plastic windows, and I had to squint as I climbed onto the stage and looked down at the twenty or so pairs of sunglasses watching me. There was no microphone or lectern to hide behind, so I was forced to just stand there completely exposed. I cleared my throat.

"Thank you all for coming," I said, reading directly from the top card in my hand. "I'm Trevor Barnett, the executive vice president of strategic planning for Terra."

On the bright side, my title kept getting weightier.

"Following an extensive review of the surgeon general's report on smoking, the major tobacco companies have determined that the report is overwhelmingly sound and that its conclusions are authoritative. In English—it means that, for the most part, we agree with it."

As it had at the hearing, the volume in the room shot up, this time accompanied by almost everyone's hands. Trainer's performance earlier that day hadn't been televised yet, and it seemed likely that these reporters were hearing the industry's flip-flop for the first time.

"Based on this report and the other statements made by the government," I continued, "we can only conclude that the executive and legislative branches are as strongly opposed to smoking as the judicial branch."

Curious as to where this was going, I peeled off the humidity-dampened top card and exposed the next one. "There is little question now that cigarette smoking can be associated with a number of illnesses. It is the position of the tobacco industry that many things are bad for you—drinking, owning a gun, eating poorly, not getting enough exercise, driving a car—but that as Americans, we should have the right to choose to do these things and suffer the consequences."

Next card.

"Based on the current environment, though, it seems that this is an unpopular view and that the American people want the government and the courts to tell them what they can and cannot do in the privacy of their own homes. We believe that this is a very dangerous road to go down and signals the beginning of the end for our great country—which was founded on the concepts of self-determination, personal responsibility, and freedom."

The next card was the last, and I stared silently at it for what must have seemed like a long time to everyone else in the tent. There was a drop of sweat dangling from my nose and I wiped it away with a shaking hand.

"In response to the obvious wishes of America's government and citizens, the tobacco companies have . . ." I lost my voice for a moment and had to clear my throat again. "Have closed their manufacturing facilities and recalled their products from wholesalers and retailers."

There were a few seconds of silence, and then everyone started shouting. I yelled over them, still reading. "We feel it's important at this point to work with Congress and the other representatives of the people to create a strategy for how—and if—this product will be sold going forward. In the interest of public safety, though, we will no longer sell tobacco products until a decision has been made on how to proceed."

When I looked up, I was surprised to see much of my audience out of their chairs and fighting their way to the tent's exit. No doubt to clean out the nearest 7-Eleven of their favorite brand.

As it turned out there was one last card. I read directly from it, though I seriously doubted anyone heard.

"I'll be answering no questions at this time."

22 I DUCKED INTO MY HOUSE, SLAMMED THE DOOR BE-
hind me, and then ran around closing curtains. When every
window had been covered, I flattened myself against a wall and put my
eye to a small gap remaining in the living-room drapes. The roving bands
of withdrawal-crazed smokers, laid-off tobacco workers, and psychotic
cigarette smugglers hadn't materialized yet, but it wouldn't be long. Staggering, pale, and skeletal, they'd soon be moving toward my home carrying makeshift clubs, sharp farm implements, and torches.

Nicotine sniffed my foot and, apparently satisfied, began rubbing up
against my leg like the cat she sometimes thought she was. I didn't pet her,
instead rechecking the curtains before finally daring to turn on a single
light.

On the flight back, Trainer had babbled even more relentlessly than he
had at dinner—dipping his shriveled toe into such subjects as diverse as
tobacco, the government, video games, flowers, and sports. Watching the
energy of a four-year-old bubble from a seventy-something's body is kind
of spooky, but I didn't let that stop me from asking a few innocuous questions designed to determine whether or not he'd gone completely nuts.
His answers proved inconclusive.

According to the light on my answering machine, my personal best of
three messages had been surpassed by twenty-six—and I'm guessing that

was only because it ran out of tape. I deleted them all without playing them, took the phone off the hook, and headed for the kitchen.

I wasn't hungry, but cooking was about the only thing that still had the power to relax me, so I whipped out my gourmet dog-treat cookbook and started in on a bone-shaped, beef-flavored snack for Nicotine. She paced furiously, seeming to fill the entire kitchen with her fuzzy bulk and forcing me to teeter around her while I worked.

Losing myself in the feel of gooey dough between my fingers turned out to be harder than normal. There was a small television hanging on the wall, and I finally gave in to temptation and flipped it on. I only had to surf about halfway through the channels before I came face-to-face with myself. Even though my television appearances so far could only be described as unfortunate, I'm embarrassed to say that it was kind of thrilling to see myself on the tiny screen.

The show cut away after "closed their manufacturing facilities and re-called their products from wholesalers and retailers," probably not want-ing to show their hardworking reporters diving for the exits. An anchor came on and confirmed that tobacco distributors were pulling their prod-uct from shelves all over the country and that the industry's plants were in the process of being locked up. Beyond that, Big Tobacco's management wasn't returning calls and Lawrence Mann, the head of the newly reor-ganized and heavily fortified Tobacco Workers Union, had refused to comment.

With nothing but ominous silence coming out of the Carolinas, the me-dia had no choice but to resort to more touchy-feely man-on-the-street in-terviews: a convenience-store worker ("They just came in and emptied out our shelves! We got nothin' to sell!"); a tobacco farmer ("I don't know what they're playing at, but you know who's going to get hurt? The hard-working people trying to provide for their families, that's who"); a non-smoker ("Hey, man, I don't know . . . It's stupid to smoke, but if you want to, go for it"). I switched the TV off before they got to the inevitable guy who'd been three hours without a nicotine fix.

The knock on the door came just as I was preheating the oven. I froze for a moment and then lunged for the light switch. Nicotine, thinking it was playtime, tried to tackle me and I barely managed to get my hands around her muzzle before she started barking.

"Shhhh!" I hissed.

The knock came again, this time louder and punctuated by a ringing of the doorbell.

I slid down the wall and wrapped my arms around Nicotine to keep her from making a break for the door. My fifty-pound advantage didn't get me far in the face of the unusual prospect of a visitor and a moment later she was free, running for the front of the house.

I followed and found her standing with her front paws on the door, barking merrily. I grabbed her collar and dragged her back, trying to decide if I should call the police.

"Trevor?"

The voice was muffled, but familiar. A trick? Withdrawal zombies could be a clever lot . . .

"Trevor? Are you in there?"

I pushed Nicotine behind me and took a deep, cleansing breath. Then, in one swift motion, I unlocked the door, yanked it open, and dragged a surprised Anne Kimball inside.

She looked frightened as I slammed the door shut—obviously unaccustomed to being pulled into dark houses by deranged men. Nicotine circled her, sniffing intently at this new addition to her world. When Anne got past her initial shock, she reached down and slid a hand across Nicotine's soft back.

"What are you doing here?"

I was kind of pissed at her—for what she'd said to me in that cafeteria, for hanging up on me when I'd (perhaps pathetically) offered her a job. For not being able to see what a sensational human being I was.

"When I said you should do something, I had no idea you'd take me so seriously," she said.

I didn't respond, instead walking past her back down the hall. The oven was probably the right temperature by now.

"What just happened, Trevor?" she said, pausing at the entrance to my kitchen long enough to get over the disorientation everyone seemed to feel at the sudden change in décor. "I don't get it."

"Hell if I know," I said, sliding a spatula under my creation and putting it in the oven. "I read a few cards that someone else wrote. That's it."

"That's not what they're saying on TV. They're saying that you're the new head of strategy at Terra and that you're behind this."

I slammed the oven shut and turned back toward her. She looked great in industrial kitchen light, too. Her jeans weren't formless like her work clothes, and she was wearing a white T-shirt that said THANK YOU FOR NOT SMOKING but also hinted at the bra beneath it.

"Really?" I said, resisting another suicidal spark of pride.

"What's going on, Trevor? They say it's actually happening—that the whole machine is shutting down."

I shrugged with calculated disinterest.

"Jesus, Trevor—when are you going to admit that you're involved in all this? You have an adjoining office to Paul Trainer, for God's sake!"

I pulled a beer from the fridge and held it out toward her. She shook her head, so I popped the top off and started in on it myself.

"Look," she continued, "even if this wasn't your idea, you're in as good a position as anyone with the company and the press to help shape things going forward. Maybe to use this as a springboard to create meaningful laws, to educate the public—"

It was wrong of me, but I laughed. In my current situation and mood, her optimism seemed a little goofy and naïve.

"Go ahead and laugh, Trevor! But you know what I'm saying is true. Maybe you'd fail and not get anything done at all. But it wouldn't kill you to try."

"Where was all this passion when I asked you to come to Terra and help me?" I said, peeking into the oven to make sure I wasn't burning Nicotine's dog biscuit.

"Come on, Trevor, you—"

"Give it up, Anne. Paul Trainer doesn't do anything that's not about profitability. And profitability is about selling more cigarettes without having to pay off the people they make sick. Like you said: The Montana thing's got them scared. And now they're doing something about it."

"But it's a desperate move, Trevor—there's no way to hide that. They're going to be losing millions a day. That could make them vulnerable . . ."

I hopped up on the counter and took another pull from my beer. "They'll get exactly what they want, Anne. They've never lost before, and they're not going to this time either."

"How can you say they've never lost before? The industry's paid out over two hundred billion dollars in the settlement with the state attorneys general, it's been forced to fund the antismoking lobby with its own money, it's been coerced into agreeing to measures to stop teen smoking—"

I laughed again, and this time she looked pretty mad about it.

"What?"

"Nothing."

"Why are you laughing at me now?"

I sat there for a moment, trying to decide what to say. She'd been brutally honest with me, and I decided that I would get my petty revenge by showing her the same courtesy.

"Come on, Anne. The settlement with the states didn't cost us a thing—we just raised prices, and in the meantime, we got nearly everything we wanted. The states agreed they could never sue us again; we tied their payments to our profits so if sales go down, they get less money. The whole thing was a tobacco industry plan to make sure the states have absolutely no incentive to actually reduce cigarette sales. For God's sake, was it completely lost on you people that our stock went through the roof after we signed that deal?" I waved around at the kitchen. "Your great victory paid for my house."

She opened her mouth to defend herself but I cut her off.

"And as far as the funding of the antitobacco lobby goes, we'd been dying to do that for years—to make the antitobacco groups reliant on us for money. And you let us tie your payments to sales, too!" I pointed at her. "Your survival is completely dependent on our holding on to our market share."

"That's—"

"And what about the earth-shattering horrors of teen smoking?" I said with anger that really wasn't aimed at her anymore. "Teen smoking rates went *down* when we were running the Joe Camel ad campaign that everyone was so freaked-out about. We just used old Joe as a bargaining chip. Here's a newsflash: Ads don't get kids to smoke! Kids smoke because it's forbidden to them and not to adults—it's a way for them to show they're grown-up. And without even breathing hard, we got you to completely ignore adult smoking, which is the foundation for teen smoking. Even better, we got you to demand huge penalties for teen smoking when every study ever done suggests the higher the penalties, the more kids smoke. That's why when you say 'fifty-dollar fine,' we say 'thousand-dollar fine'!"

She was backed against the wall now, her jaw set like a cinder block. I'd never seen those supernatural eyes of hers when there was something serious going on behind them. They were really bright.

Not bright enough to stop me, though. "There's nothing we like better than a morally outraged, holier-than-thou antitobacco crusader. You can't buy that kind publicity . . ."

"You don't think I know all this?" she pretty much screamed. "You don't think I know that half the antitobacco lobbyists out there are a bunch of political hacks who put posturing and fund-raising before results? What do you think I'm doing at SY? I'm trying to change that! I'm trying to *do* something!"

"Come on, Anne," I said, sliding off the counter. "When are you going to open your eyes and see that no one wants an end to smoking? The government gets their tax money, the lawyers get their fees, smokers get a product they love, and the antismoking lobby gets a crusade to feel superior about. Everybody's happy."

"Until they start coughing up blood."

"Must have been worth it, though, or they wouldn't have done it, right? They knew what was coming from the time they ripped the cellophane off their first pack."

"It wasn't so worth it when my mom died. She'd listened to years of the industry saying smoking wasn't dangerous and by the time she figured it out, she was addicted."

And that was the moment I should have offered to make her dinner and walked away from this fight. But for some reason I couldn't. I don't know where all the anger was coming from—thirty years of swallowing it, I guess. But I wasn't going to let her get away with the "You murdered my mom" thing.

"You *know* it's killing you, Anne. It's smoke, for God's sake! That's why when your house catches on fire they don't tell you to stand tall and take deep breaths." I pulled a cigarette from my pocket, lit it, and held it out. "Ever try one?"

She grimaced. "No!"

"Now's your chance, Anne. Pretend you're living in nineteen fifty and everybody's telling you these are good for you." I walked over and held the cigarette a few inches from her face. "Now take a nice, deep drag and see how healthy it feels."

She gave me an angry shove and I stumbled backward, slipping on some flour and landing hard on the floor.

"You might be able to fool everyone else, Trevor. But not me. It's not working anymore, is it? All that denial and all that money you handed SY hasn't made you feel one bit better about yourself."

There was a muffled crash that sounded like it came from the front of the house, but she didn't notice.

"Anne—"

"And now you're stuck right in the middle of everything. How are you going to hide from that?"

Nicotine, who had been enjoying the show, trotted off to see what was up as I struggled to my feet. I grabbed Anne, and she fought back harder than someone her size should have been able to.

"*Shhhh!*" I whispered loudly. "We've got to get out the back! Some-one—"

But it was too late. Two men came running up the hall and before I could do anything, one had grabbed Anne around the waist and pulled her away from me. I reached out and tried to get hold of her feet as they kicked helplessly in the air.

"Sir! Are you all right?" one of the men said, stepping between us.

"Let me go!" Anne screamed.

I echoed the sentiment. "Let her go!"

To my surprise, he did.

Nicotine slipped back into the kitchen, and we all just looked at each other.

Both men were dressed in stylish suits that had obviously been tailored to fit smoothly over their well-developed muscles. Their hair was long, silky, and tied back in nearly identical ponytails—the only difference be-ing that one had light hair and the other dark. All in all, they didn't fit the image of a couple of detoxing wackos.

"Who the hell are you guys?" I finally managed to get out. "And what are you doing in my house?"

"We've been hired to protect you," Blonde said. "We heard the shout-ing and a crash . . ."

Anne took a step back toward the archway that led out of the kitchen. "You have storm troopers now? My God . . ."

Then she turned and ran out of the house, but not before giving me an-other shove that again sent me to the floor.

"Are you all right, Mr. Barnett?" the blond one said, helping me to my feet and dusting the flour off me. "We're sorry if we frightened you. We'd hoped to have a more civil introduction, but when heard all the commotion—"

"Hired by who?" I said, already knowing the answer.

"Paul Trainer."

I took a deep breath and let it out, turning off the oven and pushing past them into the hall. "I'm going to bed."

"We'd like to go over the house and talk to you about your schedule and the comings and goings of your friends."

I patted my leg, and Nicotine fell into step behind me.

"I don't have any."

23

I VIVIDLY REMEMBERED TAKING THE PHONE OFF THE hook, so the incessant ringing had to be a dream. I ignored it and it finally stopped, only to start again a few seconds later.

Eventually, I reached for it and slid it beneath the covers to hide from the powerful morning sunlight coming through the windows.

"WHAT?"

"Is this Trevor Barnett?"

I didn't recognize the voice. "Why are you calling me at this hour? Do you know what time it is?"

"Uh, yeah. It's eleven o'clock."

"Oh."

"This is Gary Vandorn. I'm with the *New York Times.*"

I recognized the name. Vandorn was an unrepentant, but not fanatical, smoker who had been pretty evenhanded with the industry. A generally reasonable guy with credibility on both sides of the fence.

"I was hoping to get a response from you to Congressman Godfrey's statement this morning."

"What statement?"

"That this is all just a ploy to put pressure on the American people and that the tobacco industry is hell-bent on shirking its responsibility to everyone it's harmed and that it's the government's role to protect all the morons out there from big, bad corporations like Terra. You know the speech."

I scrunched down a little farther beneath the covers, trying to escape what little light was managing to filter through. As consciousness started to get a firm grip on me, I started mentally replaying my fight with Anne. What the hell was wrong with me? She'd walked through my door wanting to believe that I was a knight in slightly tarnished armor, and I'd completely blown it. Thirty-two years of avoiding confrontation and telling people what they wanted to hear, and I pick *that* moment to melt down.

"Mr. Barnett? Are you still there?"

I was cursing myself for telling Anne that she was wasting her time at SY until the rest of what I said started to come back to me. Jesus . . . the stuff about her mother . . .

"Mr. Barnett?"

"What?"

"Still looking for a comment."

"You want a comment? Okay, here's a comment. I'm getting kind of tired of all this political bullshit. You know how much we make off a pack of smokes? A few lousy cents. You know how much the government makes? As much as four bucks. In fact, you could go so far as to say that the sale of cigarettes is primarily a form of taxation and the main purpose of tobacco companies is to collect those taxes. For God's sake, they've exempted us from practically every piece of public-safety legislation ever enacted. They spend millions subsidizing tobacco farmers. They divert state settlement money earmarked for smoking-reduction programs into their slush funds. And all the while they stand there on TV and tell everybody what evil bastards we are. As far as I'm concerned, Congressman Godfrey can kiss my ass."

There was a brief silence over the line.

"Can I quote you on that?"

From my position at the bottom of my bed, I could feel the industry's stock plummeting, my hope of ever getting another dime out of my trust disappearing, and my chance of keeping my rather confusing forty-five-thousand-dollar-a-year job from disintegrating. And then there were all the people out there fantasizing about shoving bamboo shoots under my fingernails . . .

"Sure. What the hell."

I slammed the phone back down and gave a little thought to my schedule. I could get up, take a shower, and go to my big new office. Or I could just lie in bed all day. The choice seemed clear.

I was in the process of reaching out to close the drapes when there was a knock at the door. I'd almost forgotten about my new bodyguards.

"Go away!"

It didn't work. I watched the two men enter, followed closely by Nicotine, who jumped up on the bed and tried to get at me with that sloppy tongue of hers.

"Mr. Trainer needs to see you," Blonde said. I tried to remember if they'd ever told me their real names.

"I'm sick. I can't go in today."

Brunette walked into the bathroom, and a moment later I heard the shower go on.

"Have some coffee. It'll make you feel better," Blonde said, handing me a cup. He just stood there staring at me until I took a sip.

"Okay, we're going to leave you to it," Brunette said, reappearing from the bathroom. "We need to be in the car in twenty minutes." His face lost what little friendliness it was able to convey. "Twenty minutes. Are we clear?"

HISTORICALLY, THE PROGRESSION of the protests taking place in front of Terra's headquarters had been pretty predictable. As the Montana litigation dragged on and the blood in the water continued to embolden the press and politicians, the demonstrations had grown in size, intensity, and duration. Not an ideal trend from an industry perspective but, in a way, comfortable in its steady slowness.

So I was completely unprepared for what we were faced with when we turned off the main road toward the Terra building. Our thirty or forty die-hard antismoking demonstrators were gone, replaced by something like two hundred angry and jittery-looking people holding signs that said things like CHOICE! and FREEDOM! We were forced to slow to a crawl as we pulled up behind a line of cars trying to get past a hastily erected police barricade.

We continued forward in fits and starts, listening to the angry, unintelligible yelling that had replaced the dopey rhymes I'd become used to. The signs bristling from the crowd were surprisingly well designed and professionally executed considering the short notice, suggesting to me the involvement of Terra's marketing department.

It took about two more minutes for us to make contact with the protestors—and I mean this literally. We had to use the car's bumper to gently push them out of the way as we continued forward. I lay down in the backseat, covering my face and hoping that no one would recognize me. Obviously, my fear wasn't completely imagined because a moment later the car's locks clicked down.

I tensed when I heard a knocking on the window, and I uncovered my face enough to see what was going on. A cop was standing at the driver's window and Blonde held up a Terra employee badge for a moment before we were waved on. It seemed like an hour before the sun filling the car faded to dim fluorescent light. I sat up and gratefully took in the relative silence and stillness of the parking garage.

Blonde pulled the vehicle to within a few feet of the elevator, and the locks went up again. "You have our cell number, right, Trevor?"

I nodded.

"Call us when you're about an hour from being ready to go home. Don't try to walk out of here on foot or accept a ride from one of your coworkers, all right? And from now on, you'll be eating in the executive dining room only—don't go out and pick something up and don't order anything in."

I didn't ask about the take-out thing. The implication seemed to be that someone might want to poison me. He was just being thorough, though, right? He didn't really think anyone would . . .

"SEEMS LIKE THINGS are starting to go our way down there."

I'd gone directly into Paul Trainer's office, as directed. He was standing on his desk looking down at the crowded sidewalks below. It was the first time I'd ever seen him not wearing a jacket, and his white dress shirt was unbuttoned at the collar with no tie. Battle dress.

"Passion!" he shouted, not looking at me. "People have been wondering what the meaning of it all is ever since our pathetic species could put together a rational thought . . ."

"The meaning of what?"

"Life," he said emphatically. "Why we were put here."

He spun around on his desktop and stood there like he was onstage. "The older I get the more I think that it's passion, plain and simple. We were put here on earth to feel alive. That's all."

I moved forward a few feet, and he looked down on me as he had on the protestors. "What do you feel, Trevor?"

I wasn't sure. But whatever it was, it felt wrong—as though someone else's emotions had invaded my body and my system was having a hard time fighting them off.

"Like a quarter of the population wants me dead," I said, finally.

"That just seems like a real 'glass is half empty' attitude, son."

"Are you suggesting that I focus on the fact that three-quarters don't?"

"That's the spirit!" He held out a hand, and I helped him down to the floor.

"How about a raise, Trevor? Would that cheer you up?"

"Excuse me?"

"How much do you want? How much do you think you're worth as the executive vice president of strategy?"

"Um . . . A hundred thousand dollars?"

Trainer looked as though he was in pain. "Jesus Christ. Remember all those people who want you dead?"

"Um, plus options and four weeks of vacation?"

"Why don't we just say two fifty a year and we'll work out a nice benefits package for you over the next couple of days. How's that sound?" He offered his hand as I tried to get my mind around the fact that I was making a quarter of a million dollars a year. I remembered my conversation with Anne the night before and how I'd told her that the tobacco industry loved funding the antitobacco crusaders—loved making them reliant on tobacco dollars. Now Trainer was trying to latch me to the same teat.

I took his hand, and he pumped it energetically. "Keep up the good work," he said. "There's more where that came from."

DESPITE ITS SIZE, only the far corner of the room was being used. That's where Senators Randal, Packer, and Wakely had my father pinned against the wall. If it had been an alley instead of Terra's richly appointed boardroom, I'd have thought they were trying to steal his watch.

The three turned threateningly as Trainer and I entered, and I tried again to mimic my boss's serene expression.

"What the hell's going on, Paul? Why weren't we told anything about this?"

Fred Randal was a peculiar-looking man with a truly grand head that perched on his blocky shoulders like a geologic feature from a Roadrunner cartoon. Today, his coloring made that analogy even more literal.

"Calm down, Fred. Why don't y'all have a seat and we'll talk."

My father skirted the senators and took the seat to Trainer's left since I'd already taken the one to his right. We both wanted to be as close as possible to his protective umbrella and as far as possible from the politicians trying to stare us down.

"Have you seen the paper today?" Trainer said, pointing to a *New York*

Times sitting in the middle of the table. No one made a move for it, so I slid it toward me.

Gary Vandorn's article started out with a little history, an overview of the industry's legal problems, and the ramifications of the Montana suit. Then it moved on to me—the industry's new strategy guru—and my run-in with Scalia on TV. Spin is an amazing thing. Suddenly I found myself transformed into the "honest, flesh-and-blood spokesman that the industry has needed for a quarter of a century." I read that part three times, ignoring the heated conversation going on between Trainer and the senators and wondering how my "kiss my ass" quote was going to look in tomorrow's paper.

I noticed a thin file beneath the newspaper and flipped it open. After scanning its contents for a moment, I slammed it closed again and shoved both it and the paper back to the center of the desk.

The folder contained copies of Vandorn's medical records, as well as a summary of his smoking habits. I refocused on what was happening in the room and tried not speculate on how that information had found its way to Terra's boardroom.

"Where is it you think this is going, Paul?" Randal said.

"To a final decision on the smoking issue. Will this industry continue in the U.S. or will we move operations to a friendly foreign country and operate solely in overseas markets?"

"What the hell are you talking about?"

"It's time we find out what the American people really want. Every year, public opinion turns more against us and the lawsuits get larger and more sophisticated. We're being bled. It's time to stand and fight. Or die."

"Jesus Christ, Paul!" Randal said, waving toward his silent colleagues. "We've supported you every step of the way. We've done everything possible to keep the government and the courts off your ass. We deserved to be consulted on this."

Trainer seemed unimpressed but remained unfailingly polite. "We appreciate your support, Fred, but it hasn't done us a whole lot of good lately, has it?"

Randal looked like he was coming to the end of his patience, and the color of his big head deepened. He slammed his palms down on the table and leaned over them toward Trainer. "You're fixing to put half the people in the South—our constituents—out of work, not to mention the retailers, wholesalers, advertisers, and everyone else who relies on tobacco to put food on their tables."

And votes in his pocket, I thought.

"There isn't a retailer in this country with a single cigarette on their shelves, and we're already hearing about packs going for as much as thirty dollars. What the hell's going to happen when people completely run out? Police departments all over the country are already gearing up for a huge increase in violent crime. Pretty soon, you're not going to have a friend left in the world, Paul."

I found myself slowly scooting back in my chair, away from Trainer, away from Randal, away from the file on the desk, and away from the sixty million people who'd just been violently separated from their drug of choice.

"Something you might not have thought about, Paul," my father said, trying to end the staring contest going on between Trainer and Randal, "is that we could be liable for any violence people want to argue was the result of not being able to purchase tobacco products: domestic, workplace, road rage—you name it. They'll take the position that we knowingly addicted people to cigarettes and then cut them off for our own purposes."

Trainer shrugged. "Then we'll have five thousand suits against us instead of three thousand."

"Goddamnit, Paul!" Randal shouted. "You can't just drop a bomb like this! Not now!"

"I can't think of a better time."

I'm not a particularly political person—in fact, I've never voted in my life. So it hadn't registered with me that elections were coming up. Combine that with the sluggish economy and the government's reliance on tobacco money, and there were going to be a lot of people with their backs against the wall. A lot of very dangerous, very powerful people with their backs against the wall.

Senator Packer, a surprisingly soft-spoken and moderate man (for a southern politician), put a hand on his colleague's shoulder and gently pulled him back from the table.

"Look, Paul," he said. "I think we all agree that we're not going to resolve this in the next couple of hours. What we need to do, though, is get some of our people back to work. What about exports? That doesn't have anything to do with your legal problems here. Why not keep the manufacturing and distribution machine going for exportation?"

Trainer didn't say anything but obviously wasn't interested. Politicians were in the business of compromise, but CEOs weren't. I'd spent probably too much time thinking about all the people cut off from their cigarettes, but I hadn't yet sat down and considered the incredible economic and political impact of all this. The scale of it was almost too large to for me to comprehend.

The door to the boardroom swung open, and I wouldn't have been surprised to see the president of the United States, an angry mob, or an assassination squad walk through. I was surprised, though, to see a politely smiling Anne Kimball sashay in with a tray of sugar-coated doughnuts and coffee.

Trainer stood, as was his custom when a lady entered the room. "This is Anne Kimball, Trevor's new assistant," he said. Everyone in the room was shamed into mumbling a polite greeting.

She slid the tray onto the table and looked directly at me. "Your favorite."

I just sat there like an idiot, trying to see if it was possible to let my mouth hang open any wider.

She turned to Trainer. "Is there anything else I can get for you?"

"No thank you, Anne. I think we can manage"

I watched her disappear through the door as Trainer poured himself a cup of coffee and helped himself to a doughnut.

"The smokers are organizing pretty fast," he said. "Including a bunch of actors and musicians who've been receptive to the freedom theme we're going to go with."

Randal tried to say something, but Trainer held up a sugary hand.

"The antismoking lobby has pretty much blown itself apart over the last twenty-four hours—we've made it clear that we're not going to make our annual settlement payment that's due next week, and they're scrambling to figure out how they're going to keep their lights on." He glanced over at me. "Also, I understand Angus Scalia's trapped in his house by a bunch of protestors calling him a fascist. A nice mouthful of his own medicine, don't you think?" He focused on Randal again. "And as for nonsmokers, well, the majority of nonsmokers don't support taking away the rights of adults. So, I'd say to you, Fred, that I'm not the only one twisting in the wind here. On this issue the political establishment has gone from not having an enemy in the world to not having a friend in the world. I think you'll find it's an uncomfortable position to be in."

"Is that a threat, Paul?"

"Not at all. This is going to be a difficult time for all of us, but I want to stress that we're in it together and that all we're looking for is an equitable and permanent solution. We want people to understand the dangers of smoking and to take responsibility for their actions. It's not too much to ask."

"Reopen the factories and get distribution moving again, Paul," Randal said. "You've woken the people in Washington up. Let us go and see what we can do."

"I think they can be even more awake. We're going to force everyone to take a public position on this, Fred. No more political bullshit—people are either going to be with us or against us."

"Congress would've strung you up twenty years ago if it weren't for us!" Randal shouted. "No one's in your corner anymore, Paul. Not smokers, not nonsmokers, not your employees, not Congress, not the president. We're all you've got left—all that's standing between you and them."

"The decision's been made."

"Don't fuck with us, Trainer!"

Randal screamed this—a piercing whine that I'm guessing could be heard throughout the floor.

When Trainer's only reaction was an irritated frown, the three senators just turned and stalked out the door.

"Trevor," Trainer said when they were gone. "Could you have somebody send each of them a vanload of cigarettes? I'm guessing that before long a few smokes are going to go a long way toward getting things done."

"Should I include some nylons and chewing gum?"

He laughed and actually slapped his knee. "Son, you crack me up."

BY THE TIME I emerged from the boardroom, Anne had pretty much settled into the complex system of desks, file cabinets, and computers outside my office. There was a little pillow tied to the chair for extra lumbar support, a Boston Celtics coffee mug, and a tasteful but obvious THANK YOU FOR NOT SMOKING sign.

"What are you doing, Anne?"

"You offered me a job. The position's still open isn't it?"

"I don't know. Why do you want it?"

"Because you asked for help, and I was wrong to say no. Because maybe I don't have as many answers as I think I do. Because maybe there *are* some things to accomplish here."

"I don't know if I believe you."

"That's smart. A man in your position should be careful who he trusts. Now, do I get the job or not?"

I still wasn't convinced that she wasn't there to sabotage me for the things I'd said the night before, but she seemed uniquely qualified for the job for one compelling reason. I was still in love with her.

"I honestly don't know how much the position pays," I said, starting toward my office. "I—"

"Trying to make me even more the tobacco industry's bitch?"

I resisted looking around to see if anyone had heard, and turned back to her instead. "Come on, Anne. I never said that. And the things I did say, I—"

"I've got some money saved. I don't want anything."

I walked back up to her desk so I could speak more quietly. "Come on, Anne, you can't blow through your life savings over this. If you don't want Terra's money, why don't you just call this undercover work and stay on SY's payroll? I could—"

"I got laid off."

"Laid off? Why?"

"What do you mean 'why'? Because of you, Trevor. The tobacco companies waited until the last possible moment to tell us they weren't going to make the settlement payment they promised. John's having to cut loose nearly everyone."

24 THERE WAS ALMOST NOTHING IN MY NEW OFFICE that was mine. I'd tried sitting on the former resident demon's couch, then behind the former resident demon's desk, then on a chair by the former resident demon's fireplace, but I just couldn't get comfortable. Whether it was because I felt like he was watching me from wherever people like him—like us—went after we died or because of the brief, malevolent glares I'd gotten from the Senators Three, or if it was the way the press was so tightly linking me to this thing, I wasn't sure.

I ended up sitting on the thick carpet, leaning my back against an enormous potted tree and staring out the floor-to-ceiling windows. From my position, I could only see hazy gray sky, and I let the illusion that there was nothing else take over for a while.

I honestly don't know how long I sat like that. I would have liked to have had a way to keep myself busy, but I still didn't know what it was I did. Exactly what Paul Trainer said, I guess.

"Hiding? Smart."

I jumped to my feet, miraculously avoiding getting tangled in the tree's delicate branches, and watched my father close the door behind him. He walked over to a sideboard and poured himself a drink from a decanter I'd never noticed before.

"What do you want?" I said.

"I'm Terra's general counsel and you're the EVP-strategy. I'm here to talk about the future of the company. We do things like that on this floor."

He sat down in the conversation pit and I did the same, though reluctantly. It seemed unlikely that my father had any interest in my opinions or analyses. But for some reason, I had access to Paul Trainer and that made me a force to be reckoned with.

"What did you think of that meeting, Trevor?"

"I don't know. It would have been nice if it had gone smoother. But under the circumstances, I'm not sure it could have."

My father displayed an easygoing smile that he must have only just learned. "Maybe if Paul had let those guys know what he was going to do instead of lumping them in with everyone else and blindsiding them . . ."

I nodded. "On the other hand, if they'd known and the press found out, they could have been painted as having something to do with it. This way they're clean."

My father seemed confused for a moment. He'd never noticed that I occasionally said things that weren't completely stupid because he'd spent his life either talking over me or formulating his next sentence while I was speaking.

"Randal called me from his car, Trevor. He's coming unglued and so are the others. I'm not just talking about Packer and Wakely, here. I'm talking about pretty much every elected official in America."

"I don't think Paul expected them to be happy."

My father looked down at his glass and turned it thoughtfully in his hand. "Be careful, son. I know that Paul's a very charismatic man . . . But he's also incredibly arrogant. And sometimes that arrogance gets in the way of his judgment. I guess what I'm saying is, don't put too much faith in him."

Son?

I was trying as hard as I could to convince myself that my father was concerned about me, but I couldn't quite make the leap. It was ironic that he'd spent most of his life ignoring me but remained one of the most important formative force in my life. Somewhere down the line, I'd recognized that I only got attention from him when I failed—at sports, at school, at love—whatever. And because of that, I'd stopped trying. It had proven to be a hard habit to break.

The truth that I'd never had any reason to face was that Dad wanted me to fail and encouraged it at every opportunity. Anything he could do to keep me from replacing Grandad's big, handsome, dead jocks, he'd do. Of course, Grandad had joined the sons he'd loved so much years ago but, like I said, habits are hard to break.

"Paul's the CEO," I said. "What he does or doesn't do isn't really any of my business."

My father laughed—a heaving of his chest with no real sound. "It has *everything* to do with you, Trevor. All this happened right after you took the EVP job. *Your* face is the one on TV every night . . ."

I shrugged.

"Wake up, son. Everyone in the country thinks this is your doing."

His tone was calculated to make it clear that he considered the idea that I could in any way be involved in the orchestration of something like this absurd. Which, of course, it was.

"It doesn't matter if it's true or not," he continued. "It's perception that counts. There's been talk over the last year about him getting old, losing his edge. Maybe even getting a little senile. Now it's looking to a lot of people like he's come under the sway of a younger man. And frankly, he's not doing anything to stop that kind of talk."

"What are you trying to say, Dad?"

"Why pick you to make the announcement?" he said. "Because you're a good-looking, likeable guy who doesn't come off as polished. Because you got some good press about being honest and unguarded in that disaster of a debate you had with Scalia. But mainly because he wants someone else's face on this thing."

Nothing my father was saying was untrue. The idea that Paul Trainer had my best interests at heart was no less laughable than the idea that my father had my best interests at heart. It was getting harder and harder to forget, though, that I was the executive vice president of one of the largest companies in the world. Me.

"Right now, every smoker in America is connecting you with the fact that they can't have a cigarette and every politician in America is connecting you with the fact that a quarter of their voters are going to be going through withdrawal come election day." He continued to spin his glass in his hand. "I know that after working downstairs, this must be amazing for you. But Trainer's just using you. And you're going to end up the worse for it."

"Am I?"

"Look, Trevor, the government isn't going to put up with this. Politicians are a vindictive bunch, and they can't be blackmailed as easily as you think. When they all get together on one side, they're unstoppable. They can make you and Trainer into the most hated people in the world; they can trump up a way to put you in jail; they can dig up every embarrassing thing you've ever done . . ."

I wasn't sure I agreed. With the kind of media scrutiny we were under, it would be hard for them to trump anything up, and one of the benefits of having not done much with my life was that I hadn't really done anything all that bad. This wasn't something as simple as a UPS strike over Christmas. The government couldn't just step in, get everyone back to work, and save the day. Most of them were on record as being against smoking and had grandstanded hard and often about the evils of tobacco. It seemed far more likely that the government's approach would be to remain paralyzed and hope the industry lost its resolve in the next week or two.

"The smokers are going to turn against us, Trevor, and so are the people who are out of work or losing money because of this. The government will paint us as divisive and heartless at every turn, and now we can't even count on the local politicians to do anything to help us."

Maybe, I thought. But we had a fairly powerful spin machine, too.

"What about the suits?" I asked, deciding it was time for a change of subject.

There was a flash of anger that someone who hadn't grown up with my father would have completely missed. Obviously, the EVP-strategy questioning the general counsel *wasn't* something they did on this floor.

"We were managing those and keeping stock from crashing—something I can't say about your strategy."

There was no denying that every time I opened my mouth hundreds of millions of dollars disappeared.

"The board is going along with this in hopes of a long-term solution but they're scared, Trevor. They've lost a great deal of money. And so have we."

I hadn't had the heart to look at the value of my trust since I'd made the announcement. My life was weird and depressing enough right now without going looking for bad news.

"What do you want from me, Dad?"

"I want you to think about what you stand to gain and what you stand to lose from all this. I want you to think about whether it's worth the risk you're taking."

I RETREATED beneath the tree again after my father left, this time with the decanter. After two full glasses of scotch, I didn't feel any more settled. In fact, I felt worse.

The bottom line was that, whatever his motivation, Dad was right. What was I getting out of this? Another couple of hundred grand a year, of course, but that was pretty much theoretical. Realistically, how long was I going to last in this job? Long enough to get my hands on one or two swollen paychecks before, one way or another, I was out?

"Trevor? Are you in here? It's seven o'clock."

I peeked through the branches and saw Anne standing in the doorway.

"What are you doing down there?" she said, closing the door behind her.

"Thinking."

"'Bout what?" She pulled the chair out from behind my desk and sat, looking down at me.

"Why are you here, Anne? Seriously, tell me."

"I already did."

"This industry's older than the country. You can't destroy it from the inside any more than you can from the outside. It's immortal. Inevitable. Omnipotent."

She leaned forward and tapped my shoulder with fingers free of tobacco stains. "We'll see."

25 I DUCKED DOWN BEHIND THE SEATS AS WE ROUNDED the corner, contorting myself into a position that was becoming a little too familiar. My new bodyguards managed to ignore nearly all the subdivision's traffic laws at once as they scanned the quiet streets for any reporters or zombies who might be lying in wait. I'd seen on the news that the makers of nicotine patches were running three shifts a day to keep up with demand. Hopefully that would keep the relevant quarter of the population too blitzed to plan an effective ambush.

And what about my own smoking habit? Not surprisingly, cartons were still plentiful on the executive floor, which I now accessed from the security of a richly appointed private elevator. Honestly, though, it was more of an odd compulsion for me than a habit. I'd never smoked more than a half a pack a day in my life and now I was down to about a third of that. Obviously, I had some catching up to do.

Brunette jumped out of the suicide seat and monitored the opening of my garage from the lawn. Blonde didn't flip the locks up until the garage door was safely closed behind us.

"Where's Nicotine?" I said as we mounted what felt like a commando raid on my mudroom. She always greeted me at the door.

"Second floor."

I tried to turn toward the stairs, but Brunette stuck a thick arm out and blocked my path. I found myself being herded down the hall, sandwiched between them.

"Uh, guys, I want to—"

"*Shhhh,*" Blonde hissed as he pushed open the door to my den and entered. I felt a little push from Brunette and followed obediently.

"'Evening, Trevor."

Paul Trainer was sitting in my chair wearing one of those blue masks that you fill with hot water for a purpose that I couldn't remember. His head was tilted way back, as though he was concentrating on the ceiling, and his hands were folded neatly in his lap. A weathered-looking man with a utilitarian haircut and a macho build similar to my bodyguards' was sitting on my desk.

"How was the rest of your day?" Trainer lifted his head off the back of the chair and focused on me through the puffy slits in the mask. He looked like one of those aging actors who opened shopping malls dressed like the Lone Ranger.

"It was fine," I said, suddenly wondering if he'd overheard my conversation with my father. I tried to replay it in my mind and concluded that I had been characteristically noncommittal. My father had been the offender.

Blonde and Brunette took up positions behind Trainer, suddenly looking at me as though I was a potential threat.

"Why do you always look so goddamn worried? Have you ever had a moment's fun in your whole life?"

He seemed to want an answer.

"I don't know, Paul."

"Jesus Christ! I just gave you a huge raise and made you the number-two man at one of the biggest companies in the world. Over the next couple of weeks we're going to kick every hypocritical son of a bitch in the country right in the nuts and probably get your trust paying out again in the process. What the hell more do you want? A get-into-heaven-free card?"

It occurred to me that if you reversed the spin on that speech, it would sound more like this: *Jesus Christ! I just gave you enough money to enslave three people twice as smart as you, I've made you the figurehead for corporate evil, and over the next couple of weeks I'm going to hide behind you while we figure out whether or not this is a suicide mission. What more do you want? A get-out-of-hell-free card?*

"There you go again!" Trainer said, pointing to the expression I hadn't had the good sense to hide. "Are you thinking that you're never going to get your face painted on the side of a church? Never concern yourself with things that don't matter, son. That's lesson one."

He jumped up with the creaking bones and boundless energy of a nursing home escapee on amphetamines.

"Can you feel them out there?" He said, grabbing me by the shoulders and spinning me around toward the curtained window. "They're panicking."

"Who?"

"All of 'em. Those assholes we met with today are terrified that they're going to get voted out come Election Day. Our employees are terrified that they're not going to be able to feed their kids. The state governors are terrified that they aren't going to be able to pay their schoolteachers without their tax and settlement dollars. The antitobacco lobby are terrified that they're going to have to close their doors and find real jobs. The Fed chairman's terrified that pulling the plug on cigarettes is going to push us into a recession. And you did that, Trevor. *You* made it happen."

He squeezed my shoulders a little tighter in those cold, bony hands of his. If it was true that power is the best aphrodisiac, he was going to start humping my leg any minute.

"Do you know what the president is saying?"

"President of what?" I said, stupidly.

"Of the United States, boy!"

"No."

"Nothing! He's not saying anything. You've paralyzed him with fear.

And that fat bastard Angus Scalia is hiding under his bed while he gets picketed and smeared all over the television."

I squinted, trying to visualize the cowering hordes just on the other side of my curtains.

"So you agree with Senator Randal?" I said when it became clear that my mind couldn't conjure up the image.

"What?" he said, releasing me.

I turned toward him. "You agree with Senator Randal that we don't have a friend in the world. As your head of strategy, I have to wonder if this is a good one. I assume you're trying to strong-arm the government into outlawing suits against the tobacco industry, but I can't think of any-one who really benefits from that—even you. It seems unlikely that the suits will put us out of business before it's time for you to retire—Montana or no Montana."

Trainer glanced at my bodyguards. "I told you he was a smart kid; that we had to keep him safe. You dive in front of a bullet for this one."

Their nods were depressingly noncommittal.

Trainer sat down in my chair again, a good part of his energy spent. "Hell, you got it all backwards. *Everybody* benefits. Our shareholders and employees get a solid industry and the respect they deserve, the govern-ment gets to put this issue behind it once and for all and continue to collect billions in taxes from us. The court system stops getting clogged up with nonsense. Smokers continue to be free to make choices about their lives without government interference, the zealots get to keep their crusade . . ."

"Some of those benefits seem a little suspect, Paul. And a lot of them have kind of a long-term feel. People work in the short term—they tend to see what's in front of them, not what's over the next hill."

He nodded thoughtfully. "That you're right about. We need to make people see that this is the right thing. We need to start making some friends."

"Easier said than done."

He crossed his legs and smiled. "Maybe. Maybe not."

26 THE SUN HAD JUST SET AND THE DEEP RED OF THE horizon dramatically backlit the crane jutting up over Darius's house. The mannequin and sandbag hanging from it suggested that it was still in testing but would soon be fully operational.

I jogged across the damp lawn with my swimsuit in my back pocket, feeling the superficial calm I'd been seeking start to take hold. What I needed right now was to inject a little familiarity back into my life, and this seemed like the place to do it. My relationship with Darius was, perhaps sadly, the most stable in my life. He and I had been friends for twenty-three years now, and I'd spent more of my sentient life with him than with any other single human being in the world.

So I'd decided it would be best to spend this typically hot evening striking out with women, drinking piña coladas, and listening to Darius's undoubtedly unique perspective on the strange predicament I suddenly found myself in. It was my experience that after a few umbrella drinks, there was no problem that didn't briefly appear to be solved.

When I stepped through the open doors of the house and into the enormous room at its center, something felt different. What it was, though, I couldn't immediately identify. Light was still provided by the glare of a disco ball, music was still too loud and bass too chest thumbing, revelers seemed typically sweaty, frantic, and off balance . . .

I skirted along the wall, giving the chaos on the dance floor a wide berth as I headed for the rear exit. About halfway through, I found my path blocked by a small, oddly motionless group of people who appeared to be concentrating on some unseen object. Curious and still nagged by what it was that felt so unfamiliar, I spied on them as I squeezed past.

It was a cigarette.

I slowed some, but no one would meet my eye. Obviously, they were afraid that I'd ask to join them in passing and carefully dragging on that sad, lonely little butt.

I looked up at a ceiling that had always been obscured by a heavy haze of smoke and took a deep breath of the relatively clean air. It was the first time I'd actually witnessed a tangible effect of the things Paul Trainer (and I?) had done. Kind of weird.

OUTSIDE, TWENTY or thirty people were milling around in the light of the fading sky, drinking and talking. I weaved through them to the pool house, stripped off my clothes, stuffed them in my personal locker, and headed back out in my swimsuit.

I went straight for the thatch-and-bamboo self-service bar Darius recently had built and ducked under the low-hanging roof. There were two pretty girls in bikinis sitting in front of it, talking and drinking strawberry margaritas. They looked good, so I filled a stainless-steel blender with ice and strawberries and hit the Crush button. When I glanced up, both girls were staring at me. I couldn't remember meeting either one, but that wasn't a surprise—Darius's company seemed to have an entire division dedicated to hiring and firing girls just like these.

"Can I freshen those?" I said, pointing to their glasses. The sound of my voice broke them from their daze.

"Uh, no. No. Thanks. We're good."

I shrugged and grabbed a bottle of Cuervo, hesitating for a moment before tilting the bottle over the blender. Would I be better off tomorrow with a clear head or deadened senses? I decided to start down the path of deadened senses and see how things went.

"You sure?" I said, pouring.

The girls responded by getting up and retreating toward the house. I watched their smooth, thin bodies recede, a little confused. Because of what I looked like, most women's default position was to show at least a glimmer of interest in casual conversation.

Despite my complete sobriety at this point, it took another few minutes of mixing fruit, ice, and booze for it to occur to me that they probably recognized me from TV. Apparently, America's obsession with celebrity didn't extend to radical tobacco industry spokesmen.

I tested, sipped, and modified for a few more minutes, making a margarita of such perfection that those girls would forever regret not sharing it with me. I was carefully pouring my masterpiece into a plastic cup when Darius walked up.

"What are you doing here, Trevor? Figured you'd be too busy to slum with us."

His billowy linen shirt was half untucked from expensive slacks, and his hair fell a little too haphazardly around his shoulders.

"Your fly's unzipped."

He looked down, clearly irritated, and remedied the situation.

"Don't tell me you need a drink, Darius. I thought you were completely self-contained up there."

Nearly the entire top floor of Sinsimian was devoted to a large bedroom constructed solely for the entertainment of young ladies. To my knowledge no male other than Darius had so much as laid eyes on it since its completion, though not for lack of trying. Back before I lost interest in trying to figure out a way to break in, the door had been protected by a seemingly foolproof punch-pad-activated lock. Rumor had it—and I had no reason to doubt the veracity of these accounts—it was now protected by a retinal scanner obtained from a security company that did work for the CIA.

"Martini, right? Shaken not stirred?"

"What the hell's going on, Trevor? What's this crap with the cigarette supply?"

I rolled my eyes.

"Look, Trev, I know you don't have the juice to do something like this but the TV and papers are saying you're the man. I mean, where did all this come from? I thought you were a fucking file clerk."

I stared down at my margarita and used my finger to jab at an ice cube that had escaped the blender's blades.

"I honestly have no idea what's going on, man. The whole thing kind of came from outer space. I mean, I thought I was getting fired. I went into this board meeting and—"

"Let's walk," he said, stepping around the bar and throwing an arm around me.

"You sure you don't want anything?" I asked as he led me out onto the concrete deck surrounding the pool.

"No, man. I'm good."

"Looks like you've almost got the crane working," I said, pointing. He continued to concentrate on the ground.

"Yeah. Look, do you have any idea how many people smoke? I mean, how many times have you partied here? I'll bet eighty percent of the chicks who come here like to smoke when they drink."

I nodded and watched a girl—obviously a first timer—trying to figure out the best way to get into Darius's star-shaped pool.

"They're kind of freaked, Trevor."

I grinned and shook my head playfully. "Afraid one of your girlfriends is going to go postal and shoot up your office?"

"That's not funny, man." The anger in his voice surprised me. For a moment, I thought he was just screwing with me.

"When chicks can't get what they want, they don't have as good a time, you know?"

"I guess," I said, noticing that our trajectory was taking us toward the pool house.

"And if they're not happy, they're not as fun."

In hindsight, the fact that he wasn't speaking in his trademark three-word sentences should have told me that this was a deadly serious issue. In my defense, though, we'd been friends for twenty-three years. Let me repeat that. *Twenty-three years.* He'd known the girls at this party for something like ten minutes.

Darius jerked to a stop and since his arm was still around me, I did too.

"The thing is, Trevor, you're all over the fucking TV." I opened my mouth to protest, but he held a hand up and silenced me. "Now, I know you aren't responsible for this stuff, but it's kind of hard to explain to the girls, you know? I mean, not only have they had their smokes taken away, but now they've got the guy who did it walking around with a daiquiri rubbing their faces in it."

"This is a margarita . . ."

We started moving again, and I found myself being led into the pool house. We didn't stop until we were in front of my locker.

"Look, man, I'm thinking that maybe you shouldn't hang around here so much until this thing shakes out, okay?"

Now, I know I should have seen that statement coming. I mean, we were standing in front of my locker for God's sake. But—and I don't want to belabor the point—we'd been friends for *twenty-three years*. So, I was still a little confused as to his point, even though it would have been clear to a head of cabbage.

"I don't really hang around here much," I said slowly. "Maybe a couple of times a month . . ."

"Yeah," he said, opening my locker and grabbing my clothes.

I then found myself being pulled back outside and toward the gate that led to his front yard. Finally, my mind started to process what was happening and I stopped short.

"Darius. Are you kicking me out of your house?"

"For Christ's sake, Trevor. Don't be so goddamn melodramatic. You're freaking people out right now, that's all. Just let things cool down for a little while. I've got a company to run, and people come here to blow off steam. That's what keeps them happy and working, right?"

I couldn't remember him ever expressing any interest in his employees before. Other than his interest in field-testing every under-twenty-five-year-old girl on the payroll.

His arm tightened around my shoulders again, and I found myself without the strength to resist.

"Darius. I—"

We stopped again at the front gate, and his face lit up as though he'd just had a bolt of inspiration.

"Hey, man. What about hooking me up?"

"Huh?"

He threw the gate open wide. "Oh, come on. You've got access to smokes, right? Maybe you could get me some. You know, under the table. How cool would that be? This'd be the only place you could smoke 'em. That'd get everybody going again."

He shoved my clothes into my stomach and me out into the uncaring world. I just stood there, clutching my underwear in one hand and a half-empty margarita in the other.

"Hey," he said, closing the gate, "maybe we can get together next week—go out to lunch or something. I'll talk to my assistant and see what my schedule is."

"Uh, okay," I said, still standing there like a complete cretin.

He started back toward the house and back to the girl who was undoubtedly flung across the bed in his playroom, but craned his neck to look back at me before he disappeared. "Keep me posted on those smokes. You'd really be helping me out."

27 "Look, I'm serious. Pull the car over!"

This time they did.

I jumped out before the vehicle came to a complete stop and used the momentum to carry me through the dense bushes at the highway's edge. I made it about twenty yards before falling to my knees and vomiting. Profusely.

When my Egg McMuffin, large Coke, and hash browns had been completely purged, I struggled to my feet and walked another twenty feet away from the road. The farther I got from the asphalt, the cooler it was, and I could feel the sweat beginning to dry against my skin.

The margarita at Sinsimian had been only the beginning of a long and depressing evening. I'd made Blonde and Brunette stop at some empty dive bar on the way home, and there I'd gotten into some serious beer and frozen pizza. While I couldn't remember exactly what I'd thought about sitting there alone in that tattered booth, I was sure it was serious stuff. The meaning of life. Man's inhumanity to man. You get the idea.

My bodyguards, who had taken a booth with a better view of the door, finally cut me off around midnight. A few hours too late, it seemed.

I stumbled forward a few more feet and considered making a run for it. If I went far enough, I'd eventually find a hotel where I could take a freezing-cold shower, pull the curtains, crawl in bed . . .

Eventually, I began bushwhacking back toward the road, giving my breakfast a wide berth. When I hit the pavement, I waved Brunette out of the front seat and took his place, hoping that a clear view and unobstructed access to the air conditioner would help keep my nausea at a manageable level.

IN THE HALF HOUR it took to arrive at the cigarette factory that was our destination, I'd managed to pull it together a little. The frigid interior of the car had dried my face and given it a little color, and a half a pack of Life Savers had erased any evidence of our last stop.

The parking lot was full, forcing us to take a circuitous route through it, but the factory itself looked completely abandoned—all dark windows and locked doors. By the time Blonde bounced my SUV over the curb and started through the grass alongside the building, I was starting to wonder about all those cars.

I was there to meet with a team of low-level operations managers and personally evangelize Trainer's vision of big tobacco as a stable, universally popular industry. He'd painted a fairly benign picture of the day: a friendly discussion in which I would convince our liaisons to the common worker that the benefits of this strategy were well worth the few weeks of discomfort that they would have to endure. Then a nice lunch.

I figured out that I'd been duped again when we cleared the building and I saw hundreds of people milling around on the slope between a muddy riverbank and a hastily constructed, but as yet unoccupied, podium.

"What's going on?" I said, clasping the dashboard and leaning into the windshield. The crowd consisted mostly of men wearing baseball caps with macho logos and women trying to keep control of roving children. It suddenly seemed certain that these were the people who actually worked in this factory. Or, more precisely, who *had* worked in this factory, before Trainer and I laid them off.

"Turn the car around!"

My bodyguards pretended not to hear.

"I'm serious! I didn't agree to this! Turn the car around!"

I felt a hand on my shoulder, and Brunette leaned up between the seats. "Relax, Trevor. We've got your back."

I shoved his hand away. "I don't care if you have my back! What the hell does that even mean? Turn the car around!"

We stopped, and for a moment I thought I'd talked some sense into them. Then Brunette got out and opened my door before I could snap the lock back down. I pointed to a man running toward us, expecting them to tackle him and twist him up in some horribly uncomfortable position, but they just let him by.

"Mr. Barnett! I'm Ken Ewing, the manager of this facility. It's wonderful to meet you! I'm so glad you could find the time to come here personally."

I shook his hand but didn't get out of the car. If Ewing thought it was odd when Brunette dragged me from the vehicle, he didn't show it.

"Ken, I was told that I was going to meet with you and a few of your colleagues," I said as we were marched toward the podium. "Then we were going to have lunch . . ."

"I know, sir. I thought the same thing. Word came in just yesterday morning of the change."

"Yesterday morning?" I said. "From who?"

"It came directly from Mr. Trainer's office. He said that you wanted to talk to everyone—that we were all in this together. Management and labor alike."

There was no more than fifty feet separating us from the front edge of the crowd, and I tried not to look. "I'm not sure this is such—"

"Apparently, there were some last-minute changes to your speech and it's being faxed over," Ewing interrupted. "We should have it any minute."

I was more or less pushed up the stairs onto the podium and Blonde stayed within grabbing distance of me while Brunette stood on the grass below, scanning the approaching crowd through dark sunglasses.

As I watched the loosely gathered people compact themselves into something that looked suspiciously like a mob, I couldn't help remembering a bumper sticker I saw on one of the trucks out front.

YOU CAN HAVE MY GUN WHEN YOU PRY IT FROM MY COLD, DEAD FINGERS.

"You know ... It's Ken, right?" I said, starting to back toward the stairs, my car, and safety. "Ken, I think Mr. Trainer made a mistake." Blonde moved to a blocking position with a self-conscious casualness that proved I was a prisoner.

"There she is!" Ewing said, pointing to a woman rushing from the empty manufacturing plant. She was a little out of breath when she reached us but managed to hold out the thin stack of paper in her hand. Ewing backed away from it like it was radioactive and nodded in my direction.

I stepped out from the shadow of the podium's backdrop, suddenly realizing how hot it was in the afternoon sun. My watch said that we were almost an hour late, which meant my audience and their kids had been standing out in this heat even longer. That ought to put them in a good mood—just another example of management's disrespect for the working man.

I looked out over their rock-hard faces and decided that my father was probably right. Paul Trainer was trying to martyr me to the cause. Any minute now, someone out there with a mortgage, a kid in college, and a sick mother-in-law was going to take out a gun and shoot me. My mom would watch my assassination on her big-screen, high-definition TV, crinkle up her mildly sedated brow, and drop another Valium. My father would have the maid tape it, then never get around to watching it. Anne would shrug. Darius would throw a party in my honor and use his feigned grief to get chicks.

"I want to apologize for being late," I said into the microphone. "We kind of underestimated how long it would take to drive here."

An angry man with skin the texture of dried lava and an inevitable CAT Diesel Power baseball cap reached into his jacket, and I hit the deck.

No, I don't mean that I winced and backed up a step, or that I ducked my head involuntarily. I mean I threw myself to the ground behind the lectern. And from my position there, I watched him raise his bushy eyebrows and pull out a pair of sunglasses.

You can believe me when I tell you it's kind of hard to play something like that off. I stood back up, dusted off my slacks, and said, "Sorry, dropped my speech."

As a confused murmur rumbled through the crowd, it occurred to me that not a single thing I'd said so far was true. When had I become such a liar?

"Actually, I had too much to drink last night and I had to stop to throw up. They told me that I was meeting with a few people from management—I had no idea you guys were standing out here in the heat or I would have just rolled down the window." I didn't bother to look up at the reaction, instead centering the faxed speech in front of me and squinting through the glare.

"I speak for Terra's management when I tell you that we're sorry we had to spring all this on you without any warning—we wish we could have given some notice so you could have been prepared, but it just wasn't possible." I glanced up but found it made me more nervous. On the bright side, it turns out that adrenaline is good for a hangover.

"As y'all know," I read, "the government's been chipping away at the freedom of Americans for years now. It's getting hard to remember a time when there weren't a hundred people constantly running around trying to make laws to stop us from driving with our dogs in the backs of our trucks, or to block our legal right to own guns. A time when we could decide the way we wanted to live our private lives without interference of the courts or a bunch of politicians in Washington . . ."

I flipped to the second page. "The problem is that freedom costs, and we need to find out if America is still willing to pay the high price of it. This country is at an important crossroads. Is the government going to watch everything we do? Take away what they think is bad for us? Treat us like children? Where will that all stop? Will we have to fly to Canada to tell someone we think he's an asshole? To have a drink? To yell at our kids? To smoke a cigarette?"

I continued to read directly from the pages as my voice gained strength. "You're all aware of what the suit in Montana could cost us and what future suits like that could cost us. If this kind of thing continues, we're all going to be permanently out of work and unable to provide for our families. It was a hard decision, but it was the only one we could make. It was time to stand and fight."

I heard a gentle murmur and dared a glance at my audience. Surprisingly, there were a few nodding heads out there.

"Maybe America doesn't want us, and we're never going to go back to work. I don't know. But we need to find out before we're too weak to do anything for the people who keep this company running: you. We have some cash reserves and good relationships with our banks, and we're going use that to help out—because we're all in this together."

I flipped the page and felt a weak smile spread across my face. I would have bet good money that nothing could have made me smile on that particular afternoon.

"We're going to put everyone on twenty percent pay for as long as we can and we're arranging low-interest loans for our employees through our banks."

More nods. I wiped the sweat from my forehead and resolved to actually look at the people I was talking to.

"We'll be turning factories like this one into grocery stores stocked by our food divisions, and we'll be selling at ten percent below wholesale. Baby formula, diapers, and the like will be free."

I couldn't believe it, but a few people clapped. And, in turn, I had to applaud Paul Trainer. The baby formula thing was a nice touch.

"Don't get too excited," I said, ad-libbing for a moment. "We're just trying to figure out a way to get rid of all those snack cakes."

That got a pretty good laugh.

"We've also convinced our beer distributors to pitch in 'cause I think we could all use a drink right about now. Well, except for me . . ."

More applause.

"And don't worry about your friends and neighbors—all the tobacco companies are together on this. We're going to do what we can for everyone."

I flipped another page. "We also want you to know that you aren't being singled out here. By next week we'll have a fifty percent reduction in our headquarters staff. And if this drags on, it is targeted to go to seventy percent before long."

No clapping for that one, but effective just the same. Misery loves company.

"The bottom line is that we believe in this fight and we're ready to go all the way with it."

There was a disturbance near the stairs, and I saw Blonde blocking the path of Senator Fred Randal. He looked up at me with an expression that wasn't exactly pleading—his ego would never allow that. Let's just say hopeful. He must have been hiding out somewhere in the parking lot to see which way the wind was blowing.

I turned back to the crowd, letting Randal cool his heels at the edge of the podium for the moment. "We're going to get through this, but it isn't going to be easy. We know what we're offering you isn't much, but we hope that it shows you all how much we value you and appreciate everything you've done for this company—all the hard work and sacrifices."

More applause.

I was busy formulating an introduction for Randal when I flipped to the last page and found Paul Trainer's shaky scrawl across the top.

I expect you'll need this—

"Some of you've probably noticed Fred Randal roaming around here today," I read, "and I'm hoping we can convince him to come up here and say a few words. I'm sure you know how supportive he's been of this company and the people who work here."

I turned toward the stairs and leaned into the mike. "Senator? Can we steal a few minutes?"

He bounded up the steps and gave me one of those two-handed shakes, then went right for the mike. "Thanks, Trevor."

I took a few steps back and looked on with an appropriately respectful expression as he spoke.

"I guess we've all seen something like this coming for a long time, and as much as we wanted to put it off—there was never going to be a good time . . ."

Randal seemed to be warming up to his subject, so I sneaked back down the steps and toward my car with Blonde and Brunette in tow.

I still wasn't sure how I felt about any of this, but I was happy everybody was going to get a little help. And I didn't feel quite as hated by everyone. That was a good thing. And I hadn't been shot. Definitely something to be thankful for. All in all, not a day with much to recommend it, but it could have been worse.

We made it back to the truck without incident—everyone seemed content to let me fade into the considerable glow of Fred Randal, which was okay with me.

I opened the passenger-side door but before I could climb inside, we found ourselves surrounded.

"Step back!" Blonde shouted as he and his partner jumped in front of me. "Everyone away from the car!"

It was pointless, though. With one exception, each of the ten of the men around us weighed at least two-fifty and they'd moved in close enough that it would be impossible for Blonde and Brunette to grab their guns before they were beaten to death.

I fell backward, hitting my head on the top of the car and landing halfway in the passenger seat as a tall, thin man with kind of a longish crew cut strolled up to me.

"Mr. Barnett, I was wondering if we could talk for a moment."

I recognized him as Lawrence Mann—the new director of the Tobacco Workers' Union. I'd never met the man, but all the information I'd seen about him had been highly negative. Of course, that was a management perspective. Reading between the lines suggested that what worried headquarters was that he was an honest, intelligent, dedicated man who genuinely cared about the people who had elected him.

Blonde looked back at me as I wrestled myself out of the car. I nodded calmly in his direction and accepted Mann's outstretched hand before following him across the grass toward the factory building.

We stepped through a door that I'd assumed was locked and climbed a set of stairs to a small room above the factory floor. Mann closed the door behind us and we sat. He looked at me for a long time—long enough to make me feel uncomfortable—before finally speaking.

"So is everything you said true?"

It was a good question.

"I guess."

He smiled, exposing the crooked teeth of a person who had grown up poor but not the yellow teeth of a smoker. "I'm not sure what that means, Trevor. Let me rephrase. Do you personally guarantee it?"

I'd never met a union head, but this isn't what I pictured. He reminded me of a philosophy professor I'd once had more than Jimmy Hoffa.

"I don't guarantee anything. But I think it's likely. Trainer needs friends. He's got a few in the press, but what he really needs is the South."

"So he's buying our friendship with a few table scraps?"

"That seems overly harsh. I mean, they may be table scraps, but they're all we have to give."

"And if this drags on longer than you expect?"

"I don't know."

"I do. You'll declare bankruptcy, cut off our support, and then point a finger at Washington for starving American babies."

I shrugged. "Seems like a good bet."

"But you and Trainer won't starve."

Another shrug. "I'm not so sure about me, but no, Trainer isn't going to starve. Your people are going to hurt worse than upper management. That's just the way of the world."

He leaned back in his chair, lifting the two front legs off the ground. "You're not what I expected."

"I was thinking the same thing about you."

"Why now?" he asked. "Why is Paul pulling this stunt now?"

"Montana?" I said.

"That could set a dangerous precedent and maybe it would eventually bring down the industry, but by the time the shit really hits the fan, Paul will be dead or in a nursing home. Why do something this risky now when you could stick your successor with it?"

I didn't have an immediate answer, but Mann seemed content to sit there while I formulated one.

It didn't take long, but I wasn't sure if I should share it. After a little consideration, I decided I didn't owe Trainer anything. He'd blindsided me again today and as far as I was concerned, I was earning every dime of that two hundred and fifty grand.

"A rush," I said, finally.

"What?"

"I think this whole this is giving Trainer a hard-on."

Mann laughed and rolled up the sleeves of his work shirt. "I think you may have hit the nail on the head, Trevor. But in the end, I guess his motivation doesn't matter. The truth is that he's right—the industry is being bled, and it's going to die if we don't do something about it. We need a long-term solution."

"I suppose."

"So what's your advice to me, Trevor?"

"I'm not sure what to tell you, Mr. Mann . . ."

"Call me Larry."

"I'm not sure what to tell you, Larry. I figure you don't have many alternatives right now. Go along with this for a while. If nothing comes of it, you spent a couple of uncomfortable weeks and then can go back to the status quo. On the other hand, if we get protection from the government, our stock is going to go through the roof and so is our profitability. That ought to put you in a nice position next time your contract comes up."

"Are you a smoker, Trevor?"

A strange change in subject, and a hard question to answer.

"I smoke some. No, not really, I guess."

"A lot of my people are. And it might be hard for you to imagine how the prospect of not having access to cigarettes is adding to their stress."

I hadn't considered that. It never occurred to me that the people who worked at a production facility wouldn't be able to find a way to get them. Hats off to Trainer for his ruthless efficiency.

"Where are they?" I said. "The cigarettes, I mean."

"The company cracked down hard and fast. They had this factory closed and everyone out of it in less than an hour. What finished product we had, they trucked to a warehouse and put under nonunion guard."

I nodded slowly. "What if I could arrange that we use union guys as guards? You know, to try to help keep the jobs in the family."

I hesitated for a moment but then decided that this was the kind of decision executive vice presidents made.

"And if some smokes turned up missing, I doubt anybody would notice with everything that's going on. Now if they were to start getting sold or passed out to every acquaintance and second cousin twice removed, well, I imagine someone *would* notice that."

"I understand," Mann said in a way that gave me the impression that he really did and that I'd never have to give this another thought.

28 BLONDE AND BRUNETTE SEEMED TO BE IN GENER-ally dark moods after their failure to save me from Larry Mann's hordes the day before. Somehow, though, I got the impression that it was less about the fact that I could have been killed than it was the blow to their carefully maintained machismo.

As we swung around the corner and approached the crush of angry protesters outside the Terra Building, I slumped down in the backseat and reached for a pair of sunglasses I'd purchased based solely on their impressive size.

"Not today, Trevor," Brunette said. "Sit up."

"What?"

When we cleared the police barricades, the back windows started to descend.

"What the hell are you doing?"

"Sit up, Trevor!"

The sides of the slow-moving car were sliding against people who weren't quick enough to shove their way out of its path. I centered myself in the backseat and tried to cover my face without being obvious.

"Sit up!" Brunette said again, this time twisting around, snatching the glasses from my face, and jerking me straight in my seat.

"What the hell are you doing?" I shouted, trying to break his grip on the front of my shirt and throw myself to the floor. But it was too late. I could hear the rhythmic chanting of the mob degenerate into random shouts.

I'd been spotted.

The crowd closed in on us, and soon the windows bristled with arms as people reached for me. By the time a man wearing a Salem Lights tank top managed to get hold of me, I'd pretty much resigned myself to being torn limb from limb. Instead of pulling me from the car, though, he began shaking my hand profusely. A few seconds later, I was frantically shaking hands on both sides.

When we passed through the second set of barricades protecting the entrance to Terra's underground parking, the people fell away. I turned and stared through the rear window as they applauded and pumped their fists in the air.

"What the hell was that all about?" I mumbled, falling back in my seat.

"You're a popular guy," Brunette said.

As much as I wanted to believe that, it seemed like an awfully quick about-face. I couldn't help wondering if the outpouring of support was the result of genuine sentiment or the promise of a trunkful of smokes for everyone involved.

THE DELICATE ILLUSION that things might be looking up dissipated when one of the public elevators servicing the garage opened before the private one I'd been using arrived. My old friend Stan walked out with a heavy box in his hands, flanked by a few of my other former cohorts from the fifth floor. It seemed that with everything going on, the Ministry of Misdirection had become moot.

"Well, who do we have here?"

I feigned surprise. "Stan! Hey! How you doin'?"

"Not too goddamn well. I got laid off." He walked right up to me, bringing his face to within a foot of mine, as the others fanned out. I heard the doors to my SUV open behind me, but I waved Blonde and Brunette back.

"I guess you're doing okay, though, huh, Trevor? I bet you got a big raise to go along with all that money your granddaddy handed you. Yeah, I'll bet you're getting paid pretty good to put us all out of work."

His thick face was bright red, but I couldn't be sure if it was from rage or the weight of the box in his hands. The other three people standing around me seemed content to let Stan do the talking, but were obviously in agreement as to how much of an asshole I was. Apparently, they hadn't gotten the memo about my impending sainthood.

"I didn't lay you off, Stan. I—"

"Are you having fun being on TV while the rest of us are wondering how we're going to make it?" he said. "Guess what happens doesn't matter to you one way or the other. Somebody will always be there to take care of the Barnetts."

It occurred to me that there was a certain irony to this situation. Stan liked nothing more than to deride the people who sued the tobacco companies, arguing (incessantly) that they knew the risks and were just looking for someone to blame. As I saw it, this was the same thing. News reports or no news reports, Stan knew damn well that I hadn't gone from file clerk to policy Svengali over the course of a week and a half. He just needed someone to blame, and he didn't have the guts to yell at Paul Trainer.

It also wasn't lost on me that, much like the industry my ancestors had started, I didn't have a friend left in the world—not my parents, not Darius, not Stan. Not Anne. Thirty-two years of being terrified of the prospect of crossing anyone or making anyone angry had gotten me nothing. Seriously. Zip.

Well, I was done with spending all my time trying to keep people from hating me. If I was going to be so hated that I needed bodyguards, why not play the part?

"You know what you and your family can do, Trevor? You can just go to—"

"Careful," I said, cutting him off in mid-insult. "Someday this is going to be over and when it is, I may be the one deciding whether or not you're one of the people coming back." I looked around at his compatriots, every one of whom I'd known for years. "And that goes for the rest of you, too."

I don't think I'd made an honest-to-God threat since my days protecting Darius in junior high, and honestly it didn't feel that great. I must have been convincing, though, because Stan shut up.

"Y'all want to get the hell out of my way?" I said.

Suddenly everyone was in motion, trying to get away from me as fast as they could.

ON THE WAY to my office, I found Anne and the other executive assistants huddled around the cappuccino machine whispering. They fell silent when I walked by. Anne was undoubtedly pumping them for information that could be used against Terra and probably against me, but I didn't care. Really. Not at all.

I went straight to my desk and fell into my chair. I wanted to dive into an endless pile of work, to crowd everything else out of my mind. The problem with that plan was that I still didn't really know what I did. My desk remained neat, nearly empty, and completely dust-free.

"I saw you on TV again," Anne said, walking across my office settling into a chair. "You're making all kinds of friends."

It wasn't what I wanted to hear, and I frowned deeply. For the first time, I actually felt regret about hiring her. I found myself wondering if I could lure Ms. Davenport back.

"Looks like they're clearing out the building," she continued. "Essential personnel only. I guess that's us."

"I guess so."

She cocked her head a little. "Are you all right, Trevor? You seem kind of down."

"I'm fine."

She nodded but obviously didn't believe me.

"Paul Trainer wants to see you in his office."

I didn't move and neither did she.

"Have you decided where you're going with this yet, Trevor?"

"Wherever the wind takes me, I suppose."

"Hurricane," she said. "Wherever the hurricane takes you."

"EXCELLENT JOB, Trevor!" Trainer said as I walked through the door that connected our offices. "Excellent! That was the first step, and it was a hell of a big one. Have you seen the press on this? Fantastic. And the *New York Times* article—did you really say that Godfrey could kiss your ass?"

I nodded, and he slapped me on the back. "Now you're showing some spirit!"

My father was sitting on a sofa against the wall. "Congratulations, Trevor. You had those people eating out of your hand."

"They're not stupid," I said, a little testily. "They're with us for the short term, but they're watching. We'd better keep up appearances because they're going to get pretty pissed off if they see us hanging out at the country club while they're eating cat food."

My father's expression darkened but that seemed to be losing its effect on me.

"Of course, they're not stupid," Trainer said soothingly. "And we're going to take care of them as best we can. Don't you worry about that."

I sat down in a leather wing back uninvited. "I told Larry Mann that we'd transfer the security for our warehouses over to the union. I also implied that we'd look the other way if some of our inventory disappeared, as long as it was for personal use by our employees."

An expression of worry crossed Trainer's face but then disappeared so quickly that I wasn't sure if it was ever there. "We'll take care of it. Good thinking, son."

Trainer sat down, and my father continued the report he'd been making when I'd walked in.

"The third suit attempting to hold us responsible for an incident of domestic violence was filed this morning. They're arguing that we're liable for creating an addictive product and then cutting off supply. It's what I was warning you about, Paul. State governments are reporting a significant increase in this kind of petty violence. Police departments are absorbing huge overtime costs, and they're already looking into the viability of suing us to recoup those costs . . ."

Trainer waved a hand in disgust. "The government bitches incessantly about how the sale of cigarettes costs them billions, and now they're complaining that *not* selling cigarettes is going to cost them billions."

"Except that now they're actually telling the truth," I said. "The loss in tax income combined with the increased short-term costs is going to hit them pretty hard."

"There've also been reports of violence against convenience store employees from customers who think the stores are hoarding," my father continued. "We'll see suits from every clerk who gets so much as a scratch."

Trainer didn't seem concerned.

"Wholesalers and retailers haven't filed against us for loss of business yet, but if this goes on long enough I guarantee you they will. Right now, they're afraid because they see this as a short-term situation and don't want to do anything that might damage their relationship with us."

"Smart people," Trainer said. "I plan to make it my personal business to fuck every son of a bitch who turns against us. I suggest we make that quietly known."

My father nodded. "Eventually, though, they'll be looking at bankruptcy, and then those kinds of threats will lose their impact."

"We'll have to hope this won't go on that long," Trainer said.

"What's the government doing," I asked. "Has the president said anything yet?"

Trainer shook his head. "The silence from the White House is deafening."

"Have you called him?"

"Nope," he said. "And I'm not going to. I don't care how long it takes, that bastard's going to flinch first."

29 "Is that what's got you in such a snit?" Anne said. "The fact that those guys are around you all the time?"

We were cruising down the highway at exactly the speed limit, with Blonde and Brunette trailing only a few feet behind in an ominous-looking black Yukon. It had taken some serious arguing in the corporate jet, but I'd managed to convince them to let us rent our own car and drive ourselves to the makeshift Montana television studio. The truth was, I was getting sick of those guys. After the thing with Larry Mann and them putting me on display for those protesters yesterday, I was becoming increasingly certain that they were more a hindrance to my safety than an enhancement.

"That's part of it, I guess. They're just Trainer's storm troopers, you know. Their job isn't to protect me; it's to spy on me and make sure I perform whatever trick Trainer wants me to."

"And you're just figuring this out?"

She tucked a foot up beneath her and leaned back against the door of the car. I tried not to notice how pretty she was in the short green skirt and tan blouse that had, for some reason, replaced the women's-prison-warden thing she'd had going on before. I wanted to hold on to my anger and suspicion of her.

I assumed she'd refuse when I invited her to come along on this fool's errand, but she'd surprised me by accepting. Why had I even bothered to ask her? Honestly, because I just wasn't ready to lump her in with everyone else. Because I wanted—more than anything—for her to prove that she wasn't just another predator waiting for me to turn my back.

"Then what's the rest of it?" Anne said. "Are you upset that your father's angry with you for contradicting him?"

I'd told her about the meeting. Frankly, I'd told her too much.

"I don't care about that."

Every day, she seemed to look at me with more intensity. Or it might have just been my imagination. I wasn't sure.

"Have you been thinking about whose side you're on?"

I shrugged and eased past a tractor that was taking up most of the road. "I'm not on anybody's side."

"Don't you think it's time you pick teams, Trevor?" Her voice was a little hesitant, like she was weighing every word before she spoke. "You can't support both your father *and* Trainer."

"Trainer and my father both want the same thing, Anne—a strong, profitable company."

She smiled the way you would at a kid who believed in the Easter Bunny.

"What?"

"Nothing."

"What?" I repeated.

"This really isn't any of my business."

"For God's sake! What?"

She took a deep breath and let it out. "Trevor . . . Have you ever thought about what will happen if Trainer wins? If he gets the industry immunity from lawsuits?"

I shrugged. "Terracorp will make a lot more money."

"Because you sell more cigarettes?"

"No. Because we won't have to spend billions a year on lawyers and judgments."

"So you're saying that lawyers will become more or less irrelevant to the tobacco industry—no more important than they are at, say, a company that makes tennis shoes?"

"I guess."

"And what happens to your father? He's one of the most powerful men in the company. Will he continue to be?"

"You're saying that my father would put his own position over the good of the company?" My voice was a little too loud for the confines of the car. "What you're forgetting is that his trust is bigger than mine. He stands to make a huge amount of money if the industry's stock rebounds. Did you ever think of that?"

She twisted back around in her seat and looked through the windshield at the endless, rural landscape.

"I guess you know him better than me."

WE WERE broadcasting from a high-school gym, though the *Hardball* set had been flawlessly re-created and I doubted even the biggest fan would be able to tell that we weren't in the regular studio. My opponent was C. William Ivers, the lead plaintiffs' attorney in the ever-present Montana suit and the most dangerous man in the world from the industry's point of view.

"This is an incredibly obvious attempt by the cigarette industry to coerce the government into protecting it from lawsuits—to return it to a time when it had absolutely no obligation to the people it kills and to allow it to continue to peddle death with absolutely no responsibility to anyone other than its shareholders."

He looked over at me and decided to mimic Angus Scalia's fabulously successful strategy. "Trevor Barnett, here, is the figurehead for all this. A man whose family practically started the tobacco industry and who's become obscenely wealthy from the suffering of others. A man who wants to get his income back on track at any cost."

This time I was a little better prepared and a lot less nervous than I had been in my first television appearance. And while Ivers was a formidable man—taller than me, with weathered skin and a loud, charismatic way of speaking—he seemed pretty mundane compared to the hundreds of unemployed and undoubtedly well-armed tobacco workers I'd talked to a couple of days ago.

"First," I said calmly. "Let me clarify this whole issue of my personal finances. I have about a million dollars' worth of tobacco company stocks and I get a percentage of the dividends and capital gains that come off them. That income is zero right now and has been for years. I can't tell you how much I've been paid by the trust over my lifetime, because I honestly don't know exactly. What I can tell you is that it was just enough to cover college, a car, some furniture, and a down payment on a house that Mr. Ivers here wouldn't live in on a bet. Until I was recently promoted, I made just over forty thousand dollars a year, and that's what I lived on. There are a lot of investors out there who stand to gain a lot more than me if the tobacco industry stock starts climbing again, and there are a lot of people who have more to lose than I do if the industry tanks. Mr. Ivers and his colleagues would be examples of people who have the potential to lose an incredible amount of income if Tobacco is protected from suits. The lawyers in the settlement with the states have already collected over ten billion dollars."

"I make a good living," Ivers admitted. "The difference is, I do it protecting the people and not killing them."

I smiled serenely. "I once saw a lawyer on TV talking about how California was thinking about blocking frivolous suits against skateboard parks so they could be reopened and help get kids off the streets and into a supervised environment. She actually argued with a straight face that the state would stop maintaining its streets and sidewalks if these kids had parks to play in. Did she have these children's safety and best interests at heart? I don't think so. And it's the same thing here. Let's say that the tobacco industry loses the Montana suit. We don't have two hundred and fifty billion dollars, so the plaintiffs are most likely going to have to settle for a couple billion. That means Mr. Ivers and his colleagues get hundreds of millions and each plaintiff walks away with a few grand. Who's benefiting here?"

"Absolute nonsense," Ivers said. "You're just trying to divert attention from the fact that you're looking to put an industry that kills hundreds of thousands of innocent people a year above the law. We all know what the industry would do with that kind of freedom."

Chris Matthews seemed content to just sit there and listen, so I dove back in.

"Everyone out there, if they think logically about it, agrees that no one but attorneys benefit from these suits. I mean, they don't keep anyone from taking up the habit, they don't make anyone quit, they don't cure anybody of cancer, and generally the plaintiff never sees any money. So if all this legal wrestling doesn't do anyone any good—other than you— why not just put a stop to it and instead create some clearly written laws that *will* do some good?"

"I have to disagree, Mr. Barnett. The settlement with the states increased the price of cigarettes and helped put in place a number of rules that curbed your industry's ability to market to children. I'm still not sure I understand how making you answerable to no one helps anyone but you."

I considered that for a moment. "I'll concede the point that litigation has precipitated some changes in the cost and marketing of cigarettes, though I'd argue that those changes haven't done anything to reduce smoking rates. To answer your question about how making us answerable to no one might help. Maybe it'll make people think twice about smoking when they know it's their sole responsibility? Or maybe that's not what the American people decide. Maybe they decide that cigarettes should be outlawed. Or regulated as a drug? Or maybe everyone who buys cigarettes should be required to sign a document stating that they understand the health risks and that they consider those risks acceptable. Any way you slice it, it's an exciting time."

"But you're not counting on them being outlawed or regulated, are you? You're looking to use smokers' addiction to nicotine and the government's addiction to tobacco money to strong-arm this country into caving in and giving you everything you've ever wanted."

It was strange timing, but I realized that wasn't what I wanted. And while national television probably wasn't the place for it, I felt an answer to Anne's question about whose side I was on begin to form in my mind.

"We're not trying to use our position to force anyone's hand here," I said, despite the fact that I knew this was exactly Paul Trainer's plan. "I'm not suggesting that we protect the industry with some one-sided all-encompassing piece of legislation. We need to think about a comprehensive policy that would be the product of a dialog between the interested parties: us, smokers, the antitobacco lobby, and the government."

"So you're trying to get me to believe that an industry that has never done anything positive without being dragged kicking and screaming is going to make these concessions out of the goodness of its heart? Why am I not buying that?"

"Listen," I said, "there are studies that say seventy-five percent of smokers want to quit and only about three percent succeed. That should tell you two things: Smoking is not a desirable habit, and it's addictive. We're not selling cigarettes for people's health and if anyone out there thinks we are, then frankly they're idiots. We're selling them because people want them and we want to make money. Period."

"There are a lot of companies out there making money, Mr. Barnett, but they don't resort to killing people to do it. The fact that you can be so glib about this product suggests to me that you've never witnessed the horrible deaths it produces—that you just look at reports and numbers, and ignore the human reality. I suggest you go to a hospital and watch someone suffocating to death before you get on TV again."

"I want to be completely clear here—I am not suggesting that people start smoking. In fact, let me say to anyone watching this program who's considering it, I strongly recommend that you don't. Mr. Ivers is right. It can kill you, and your death probably won't be pleasant."

I couldn't help wondering if Paul Trainer was having chest pains yet.

"The question we need to answer," I continued, "isn't whether Americans *should* smoke; it's whether they should be *allowed* to smoke. Are Americans smart enough to understand the risks and benefits? And if they are, are they willing to take responsibility for their own actions? If not, we need serious legislation to completely ban this product just like we have narcotics. But having the government's executive and legislative branches saying 'smoke up' and the judicial branch saying 'stop' is ridiculous and counterproductive. We all need to get on the same page. What page that is, I don't know." I paused for a moment and mentally confirmed that what I was about to say was compatible with my new policy of complete, twenty-four-hour-a-day honesty.

"And, you'll be surprised to hear, I don't really care."

30 "I JUST DON'T GET IT," ANNE SAID. "WHAT'S PAUL Trainer's angle?"

We were back on the road again, speeding along the worn asphalt as the afternoon sun created startling contrasts in the endless rural land-scape. I watched the mesmerizing rows of crops as they sped by, almost completely ignoring the empty road ahead.

"Why would he tell you to say those things this early in the game? I mean, I know he's trying to do this honesty schtick to boost his popularity, but that seems like it went too far. What's he after?"

"I can answer that. Paul Trainer wants everyone in America to smoke a pack of cigarettes every single day of their lives, and he doesn't want to be held accountable for it."

"Only a pack a day? Why not three? It'd triple his income."

I shook my head. "Three'd kill you too young, and dead people don't buy product. We figure a pack a day is the magic number that maximizes sales over a smoker's lifetime."

"I don't think I've ever heard that statistic."

"It's not really a number we advertise."

She turned in her seat and tucked a panty-hose-wrapped leg beneath her again.

"Why wouldn't Trainer just let everybody sweat for a while? I'm not sure how suggesting a compromise so soon fits into his strategy."

I shrugged. "It probably doesn't. I just made that stuff up."

"What?"

"I just made it up."

I couldn't stop a slight smile from spreading across my face. For the first time in all this—maybe the first time in my life—I felt kind of . . . free. I was the master of my own destiny and whatever happened to me, I figured I could handle it. Temporary, groundless euphoria, maybe, but euphoria just the same.

"So what you're telling me is that you just went on national television, admitted cigarettes cause horrible deaths, and suggested that the industry would make concessions for lawsuit protection without any authority at all?"

"Well, I *am* an executive vice president, you know."

In the rearview mirror, I could see the black Yukon with Brunette behind the wheel. Blonde was in the passenger seat, leaning into the windshield pointing angrily at his cell phone. Obviously Paul Trainer had heard about my performance and wanted to have a little chat. I hadn't brought my phone, so I just sped up.

"You're serious," Anne said. "You just sat there and said whatever you felt like saying."

"Pretty much."

"Are we having a breakthrough, Trevor?"

"I think maybe I am."

"Just in time for them to throw all your stuff out on the sidewalk."

I smirked. "You'd think, but every time I figure I'm about to get canned, I get promoted."

Anne laughed and shook her head. "Not this time. Do you have any idea how hard it's going to be for Trainer to walk away from a promise his executive vice president made on national TV?"

Of course, I was still a fraud—the same hapless pawn Trainer had been pushing around since all this started. But perception was reality in this business, and Trainer had created the perception that I had power—if only to draw attention away from himself. He'd counted on the fact that

I'd just go along, and I couldn't really blame him. Until now, it had been a reasonable bet.

"So don't keep me in suspense, Trevor. Have you finally decided what you want out of all this?"

Her, of course. But I didn't think that was the answer she was looking for.

"It wasn't that big a breakthrough."

"Guess I shouldn't expect miracles. A step in the right direction, though, if you ask me."

"What about you, Anne. What do you want?"

"That's easy. I want everyone to quit smoking and live a long, happy life. I want to stop cigarettes from tearing families apart. I want to make sure that no one ever has to walk into a doctor's office and find out that the tobacco industry's killed them."

"But how far are you willing to go to get that?" I said. "Do you want to outlaw cigarettes? To take away a person's right to make that choice? Isn't the truth that those people walk into the doctor's office and find out that they've killed themselves?"

She didn't immediately respond, and I glanced over at her as I let the car glide toward a four-way stop. For the first time, she looked a little uncertain about her place in the world.

"On that subject . . . Well, let's just say I'm still waiting for *my* breakthr—"

The truck came from nowhere.

It was one of those ridiculously long, extended cab rigs with heavily tinted windows and an enormous black push-bar on the front. I slammed on the brakes, and threw an arm out to keep Anne from sliding around her seat belt.

The truck, which had run its stop sign, looked like it was going to pass harmlessly in front of us, but at the last minute it swerved and smashed into the front of our car. My head hit the steering wheel and Anne pitched sideways, squashing my hand between her and the dashboard.

I think my disorientation after the crash was more from surprise than my impact with the wheel, and I managed to shake it off pretty quickly.

"Are you all right?" I said.

"I think so. Yeah, I'm—" She suddenly went silent, staring past me toward the truck that had hit us. I craned my neck, causing a stream of blood to flow painfully into my eye, and saw four men running toward us. For a moment I thought they wanted to help, but then I saw the guns in their hands.

"Oh, shit," I mumbled and then twisted around the other way and looked through the back window. Blonde and Brunette had their doors thrown open and were partially hiding behind them, guns drawn. Thank God.

The sound of machine-gun fire is unmistakable but a lot louder than you'd think. When it started, I released my seat belt and pulled Anne down, covering her with my body. In retrospect, a kind of heroic act that I'm not sure I can really take credit for. It was strangely instinctive, and I'm still surprised that I reacted that way.

When I raised my head and took a peek out the back window, the windshield of Blonde and Brunette's Yukon was full of bullet holes. I ducked down again, waiting for them to return fire, but instead heard the squeal of tires. Another peek confirmed what I already knew. My bodyguards were flooring it down the road the way we'd just come from.

"You've got to be kidding me," I said as my car door was pulled open and someone grabbed me by the hair. I tried to keep my grip on Anne, but it wasn't possible. A moment later, I found myself lying on my back in the street with my legs still halfway in the car.

The man hovering over me, strangely distant at the other end of a rifle barrel, was dressed completely in camouflage and had a handkerchief wrapped around his face like a bandit from the old west.

I heard Anne scream and I tried to sit up, but the guy slammed the gun barrel into my chest hard enough to take my breath away. He yelled something in a language I didn't recognize, but still managed to make his point: If I moved again, I was dead.

"Trevor!"

It was Anne's voice. My vision had cleared enough to see her struggling against a similarly dressed man who had a pistol in one hand and her hair in the other. I felt a sudden wave of fury and fear, but the man above me anticipated my reaction and moved his gun so the barrel rested right between my eyes.

"She doesn't have anything to do with this," I shouted. "She's just my secretary! Let her go!"

They ignored me, of course.

The gun swung away, and the man holding it dragged me to my feet. I was probably six inches taller than him and he stepped back warily, motioning me toward the truck.

Anne was about twenty feet away, doubled over at the waist, trying uselessly to break her captor's grip on her hair.

"What do you want?" I said, my voice shaking slightly.

The guy covering me motioned with the barrel of his gun again, trying to get me moving toward the truck.

Then a strange thing happened. There was this deep, chest-rattling thud that kind of reminded me of the bass from one of those powerful stereos people put in their cars. When I looked back at Anne and saw the blood spattered all over her, my legs nearly buckled beneath me.

"Anne! Oh my Go—"

But she was fine. I shifted my gaze a bit and saw that the jacket of the man holding her was partially shredded and his chest was gushing blood. He crumpled slowly to the ground, but didn't release his grip on Anne's hair. She managed to stay on her feet and one final jerk freed her.

I don't know how long it took for all this to happen, but at about the time Anne was stumbling backward away from the dead man on the ground, I heard the echo of what sounded like a distant gunshot.

Three of our attackers were still alive, but they seemed to have forgotten I was there. They were all shouting and pointing and shooting past me up what seemed to be an empty road.

Now, that probably would have been a good time to make a run for it or dive behind the car, but I didn't. I just stood there, squinting against the glare of the setting sun and trying to see what everyone was shooting at. There *was* something . . . A car. Not the one Blonde and Brunette had been driving, though. This one was smaller. And blue. Or maybe gray. Whatever color it was, it was parked in the middle of the road a little over half a mile away.

I heard that weird thud again, and out of the corner of my eye saw the man who'd dragged me from the car pretty much explode. It wasn't like the movies, where the victim always clasped his chest, teetered awkwardly, then crumpled dramatically to the ground. Instead, a big part of the guy's chest kind of disintegrated and sprayed through what was left of his back. In less than a second, he went from a human being to a shredded piece of meat.

I felt two hands clamp down on my arm and then a relatively weak tug.

"Come on, Trevor! Run!" Anne said over the inevitable rumble of a gunshot.

I blinked at her stupidly and then let her pull me around our rental car, where we ducked down behind the tire. She wrapped her arms around me and I followed suit, pulling her close and wondering if it was her shaking or me.

"We've got to get out of here," I said as my brain slowly started to process information again. Admittedly, not a tricky deduction, but it seemed right on point.

I leaned back, still holding on to Anne, and examined the front of the car. It was totaled—if we were leaving, it wasn't going to be in this thing.

I knew that two of our attackers were still alive, though I could only see the one shooting over the hood of the giant pickup. He suddenly ducked back behind the cab and a moment later half of the front windshield, the entire driver's-side passenger window, and most of the metal between them exploded. I watched him stagger back and fall to the ground, his face mangled by flying glass.

As the shot finally sounded, the other man reappeared, dragged his compatriot into what was left of their truck, and then left us in a fog of burning rubber.

Anne watched them go by, obviously as confused as I was. She released me and pulled back a few inches. "Are they . . . Are they gone?"

I got on my hands and knees and peeked around the front of the car again. From that position, all I could see was the two mutilated bodies and the expanding pools of blood beneath them. The old cliché "eerie silence" is the only way I can describe the atmosphere at that moment. Everything had gone from complete chaos to complete stillness in a matter of seconds.

Anne tugged on my shirt, and I turned in time to see her wipe a tear from her cheek. "What do we do now?"

"I don't know. Run?"

The road was bordered by endless miles of grass cropped short by grazing cattle. Why ranching? Why couldn't we have been pinned down next to a tall, dense crop of corn? Or an impenetrable forest of pine trees?

She shook her head. "We'd never make it. What about your cell phone?"

"Didn't bring it," I said, sounding doomed even to myself.

"Do you think your bodyguards went for help?"

I didn't answer, and her eyes dulled a little as fear was replaced with resignation.

How could I have been so stupid? How could I have gotten her involved in this? If she got hurt or killed, it was my fault.

I turned and started to crawl around the front of the car, doing my best to ignore the dead man only a few feet away. Anne grabbed my foot. "What are you doing?"

"I'm going to take a look."

"No!" she said. "It's too dangerous . . . You . . ."

Her voice trailed off, probably because she realized it was only marginally more dangerous than sitting there and waiting for whatever was to come.

I crept along the smashed bumper, took a deep breath, and then eased my head out into the open.

"What do you see? Anything?"

I did see something. A lone man about a hundred yards away, running fast with a rifle in his hands. He stopped suddenly with an economy of motion that suggested mindless reflex and aimed the gun right at me. I pulled back so fast I lost my balance and landed on my back. The gunshot I expected to hear never materialized.

"What is it?" Anne said, slipping her hands beneath my arms and helping me back to the relative safety of the car's tire. "Did you—"

"You can come out!"

The accent was undoubtedly British and sounded like it was coming from just behind the car.

"Hello? Come out, please!"

I took a deep breath and slowly stood, raising my hands and moving away from where Anne was hiding. She made a grab for me, but I was already out of range.

He was thin, probably six feet tall, and had a slightly sunburned face beneath short brown hair. His chest was heaving a bit from his run, causing the rifle in his hands to rock back and forth.

"Where's the girl?" he said, scanning the rolling landscape around us.

"It doesn't matter. She's just my secretary. She doesn't have anything to do with this."

He seemed to be only half listening as he continued to concentrate on the open terrain. "Come over here."

I did as he asked, but stopped when I got to within a few feet of him.

"I don't think you understand, Mr. Barnett, I'm—"

Now, you'd think I'd have started to put some things together here, but I was surrounded by mutilated bodies, terrified that I was going to die, furious at my wasted life, and wracked with guilt about dragging Anne into this. I had a lot on my mind.

The man seemed even more surprised than I was when I dove for his rifle.

He sidestepped much faster than I'd thought he'd be able to and swung the butt of the gun in a powerful arc that was going to kill me as I stumbled by. It stopped just before it impacted my skull, though, and I ended up sprawled out on the asphalt unharmed.

"Good try," he said, sounding generally impressed. I flopped over on my back and watched him as he began moving to his right. There was a brief flash in his eyes when he spotted Anne and he jogged forward, grabbed her by the arm, and pulled her to her feet. She wasn't fighting back this time, and I could see more tears reflected in the fading sunlight.

When he released her and began pawing at her blood-covered chest and stomach, I thought it was a prelude to his raping her and I started to get to my feet. It quickly became clear, though, that he was just trying to figure out if she was injured.

"Don't cry, dear," he said, holding up a now-crimson hand. "It's not your blood."

**31** "WHAT THE FUCK IS THAT?"

We were in a small interrogation room with two detectives, one of whom seemed pretty angry.

Anne was sitting next to me wrapped in a blanket, squeezing my hand beneath the table. Stephen, the man who'd saved us, was sitting across the table from us dipping a tea bag in a steaming cup of water. All I knew about him besides his name was that he'd been hired by Terracorp to keep me safe. The police had come screaming up the highway before Anne or I could think clearly enough to formulate more complex questions.

"What the fuck is that?" the detective repeated. He was a thick man, kind of square (in shape, not personality), wearing a white dress shirt, tan pants, and cowboy boots.

There were two unloaded rifles on the table: the one that Stephen had been carrying when he'd run up to us and the one Detective McEntire was pointing to. The latter was enormous, with a black barrel about a yard long, a ridiculously high-tech scope, and a tripod thingy (you'll have to excuse my ignorance of firearms) that made it look like something a Marine would shoot tanks with.

"That's a Barrett M-95," Stephen said. "I assure you that I purchased it legally and have a valid permit."

It seemed absurd that it could be legal to own one of these things, but you'd swear from his casual manner that every man, woman, and child in America had at least one lying around the house for emergencies.

"Do you have a permit to run around our country blowing people's brains out from half a mile away?"

"That seems rather inflammatory," Stephen said evenly. "I didn't *blow anyone's brains out*. I was simply forced to engage four men who were attacking my client. In the process, I regret that two of them were killed."

McEntire leaned on the desk, supporting himself with balled fists. "'Killed,'" he repeated. "I'd say smeared all over the pavement."

Stephen didn't react to the statement.

"He saved our lives, sir," I said. "You saw our car. Those guys rammed us and started shooting with a machine gun. If it weren't for Stephen . . ." I fell silent. At the relatively young age of thirty-two it was hard to realistically think about being dead, but I was managing pretty well right now.

"They started shooting," McEntire said slowly. "This is at your other bodyguards, right?"

I nodded. We'd already been through this.

"What happened to them?"

"They drove away," I said.

"They drove away." He didn't seem to believe me. "Who are they? What are their names?"

"I'm not sure," I said honestly. "I think they told me when we first met, but I just called them Blonde and Brunette. I forgot their real names."

Stephen laughed and continued to bob his tea bag.

"Do you have a phone number?"

"Not with me," I mumbled, sounding like I was lying, even to myself. "Don't I get a lawyer or something?"

"You're not under arrest," McEntire said. "Look, I recognize you from television. I know who you are—"

I scooted back a little in my chair, and Anne gave my hand a reassuring squeeze.

"You don't smoke, do you?"

Stephen, my guardian demon, snickered some more but remained focused on getting his tea just right.

"Not anymore," McEntire said.

That answer seemed a little evasive to me.

"Look, if you know who I am, then you know that—"

"What I know is that I've got two dead bodies lying right in the middle of my jurisdiction and I'm goddamn well going to get some answers." He turned back to Stephen. "What about you. You must know these other bodyguards. Who are they?"

Stephen frowned. "Cheap muscle. Good for show and for keeping the minor crazies away. I have no idea where they are at this precise moment, but I'm sure they'll surface in the next day or so and that they'll be happy to give you a statement."

"And how are you connected to them?"

"I'm not really. I specialize in countering more organized threats. I prefer to work with a lower profile."

"Reality check, here," McEntire said. "You practically cut two people in half in the middle of the street! That doesn't seem low profile to me."

Stephen shrugged, and a moment later the only door to the room swung open. A man who was probably in his late twenties walked in holding a single sheet of paper in his hands. "We got something."

"What?" McEntire said.

"Stephen Hammond, thirty-six, British nationality. It says he served in the British Air Force for more than a decade but resigned a few years back."

"The Air Force?" McEntire said incredulously.

The younger man nodded and squinted at the page. "Yeah. It says Special Air Service."

McEntire's expression darkened and he glared at Stephen, who was now sipping his tea cautiously, apparently concerned that it might be a tad hot.

"That's not the Air Force," McEntire said, finally.

"No? What is it?"

"What it is, is a combination between a Navy SEAL and a cold-blooded assassin."

"Do you know who they were?" Anne cut in, obviously sensing the increasing tension in the room. "The people who attacked us, I mean."

"It could have been anybody," McEntire said. "Take your pick. Disgruntled smokers, laid-off cigarette industry employees . . ."

"Don't forget farmers," the younger man cut in. "I read yesterday that some people in South Carolina torched a tobacco farm and by the time the fire trucks got there, there were thirty people just standing around breathing deeply—"

"Would you mind if we didn't go through all the people who want me dead?" I said. "It's kind of depressing."

The door opened again, and this time three extremely well-dressed men rushed in.

"My clients are clearly in need of immediate medical attention! Someone call an ambulance." I recognized Daniel Alexander, the New York attorney who had been so instrumental in making me look like an ass in front of the board.

"They weren't injured," McEntire protested, trying to block the attorneys' path to us.

"You damn well better hope not," Alexander threatened.

McEntire turned toward a balding man standing in the doorway. "Captain! Look, they can have these two, but the other one's admitted pulling the trigger. I've got to at least finish getting a statement from him."

"That's absolutely ridiculous!" The volume of Alexander's voice rose to a near shout. "Mr. Hammond is a highly decorated former soldier who was hired by Terra to protect Mr. Barnett. There is absolutely no reason to believe anything but that this was a necessary act and purely in defense of innocent life—something we'll provide additional witnesses to corroborate."

"But—" McEntire started.

"And furthermore, Mr. Hammond is critical to the safety of my clients. Are you prepared to provide equally qualified protection while you have him?"

"Captain!"

"We're going to have to let them go for the time being," the man in the doorway said.

"But, sir, he just killed two men!"

"Hold on to his passport and turn 'im loose."

"Fine," Alexander agreed. "You'll have the additional witness statements by tomorrow and, of course, we'll cooperate in every way possible with your inquiry into this unfortunate event."

One of Alexander's minions grabbed me and Anne by the shoulders and helped us to our feet. No one seemed to want to get too close to Stephen, so he had to get up under his own power.

The police captain stepped aside, and we were rushed through the door and down the hall. I kept hold of Anne's hand as we passed through the station and out into the cool night.

There was a limousine waiting at the bottom of the steps and we headed for it, surrounded by our battalion of lawyers.

Anne suddenly sped up, pulling me along with her and catching up with Stephen, who was walking a few paces ahead.

"Mr. Hammond," she said, "I don't think either of us has said it yet. But thank you. Thank you for saving our lives."

He smiled but didn't look at us, instead scanning the buildings across the street. "It was my pleasure, Anne. And please, call me Stephen."

32

"I DON'T BELIEVE IT!"

"What?" Anne said. She was lying across the car's back-seat, half asleep. We'd endured a long and silent flight home after being released by the police. I felt like I'd been awake for a week, and I think Anne felt worse.

"The press," I said.

They were everywhere. The vans and cars that couldn't find a spot on my lawn or in my flower beds were lined up along what had once been my quiet street. It was only six A.M., but many of my neighbors were standing barefoot in their dew-covered grass examining the chaos I'd brought. Reporters were interviewing some of them, no doubt taking careful notes on all my failings as a neighbor and human being.

"I think you should talk to them and get it over with, don't you?" Stephen said. "I mean, if you put it off, they're not going to give up. They'll just hound you."

"Are you kidding? Look, I'll do it, but not now, okay? I can barely think straight. Tonight. I'll do it tonight."

He tapped the brake and turned slowly into what little was still available of my driveway as I slumped down in my seat. "Do me this favor, Trevor."

The breath went out of me in a long rush. Stephen was no doubt under strict orders from Paul Trainer not only to protect me, but also to get me to talk to the press while I was looking my worst and most sympathetic. What could I say? It's hard to refuse when a man who just saved your life asks a favor.

"Okay, Stephen, I'll do it," I said as the car was surrounded. "But Anne doesn't have to talk to anybody. Agreed?"

He nodded and stepped fearlessly into the microphone-wielding mob. "Everyone back," I heard him say. Miraculously, they all complied. Apparently, word of his particular talents had spread.

Anne and I stepped out into the space Stephen had provided, and I put an arm around her. "Go on in. I'll deal with the press."

Stephen shook his head. "We all go."

I did as he said, keeping my arm around Anne as we followed him through a corridor of loud, but surprisingly well-behaved reporters. A young woman with earrings that seemed too big for television tried to step in front of us, but Stephen waved her back with a courteous, but authoritative motion.

Although I was positive that I'd locked it, the front door was open. Once inside, I pushed it closed with my back, sagging against it and wishing I could crawl into bed for the next month. But there were bright lights emanating from my living room, so that probably wasn't going to happen.

"They're waiting for you," Stephen said, pointing to the back of the house.

I nodded and turned to Anne. "You okay?"

"Sure," she said quietly. "I'm fine."

I could hear barking coming from somewhere upstairs. "Could you go find Nicotine—make sure they didn't lock her in a closet or something?"

"Okay."

I slid a hand around the back of her neck and ducked down so she had to look at me. I could feel the soft strands of her hair brushing against my skin.

"You're sure you're okay?"

She forced a smile, and I reached over and pushed the Play button on my answering machine, starting the first of two messages. I had to admit to a mild sense of elation when Darius's voice filled the hallway. "Man. I just heard. Are you all right? Call me. Okay?"

I wondered how he'd gotten my new number—I'd changed it a couple of days ago and given it to almost no one. However he'd managed it, it had to have taken a little effort. Maybe he was regretting our last meeting. Realizing that our friendship—

"Oh, and hey. Let me know what's up with those smokes. Things are getting kind of ugly over here."

My shoulders drooped as what little strength I had left began to fade. I shoved my hands in my pockets and listened as the second message came on.

"Trevor! Jesus, boy. What can I tell you? The world is full of psychopaths!" Paul Trainer's voice. "I'm glad to hear you're all right. You listen to Stephen, now. He's the best. The best! That's why I hired him to keep an eye on you."

The tape beeped, and I listened to it rewind. Neither of my parents had called. Guess they were busy.

Anne pressed against me from behind and leaned around so she could see my face. For some reason, I had the impression that she knew what I was thinking.

When I got to the living room, I was rushed by a woman whose name was one of those goofy alliterations local newscasters favored, but that I couldn't remember in my present state. She was from a station in Greensboro and her reports, regularly picked up by the nationals, had been pretty fair to both me and the industry so far.

"Mr. Barnett, could you just step over here?" she said, pushing me in front of a bank of portable lights that hurt my eyes.

"Why don't we just start with you telling us what happened in your own words," she said, dispensing with pleasantries in favor of efficiency.

"Um, well, there isn't that much to tell. I was coming back from an interview in Montana and a truck rammed me and then these guys with guns jumped out."

"And they fired at you, didn't they?"

"What?" I said, having a hard time tracking on the conversation. I wanted to be upstairs with Anne and Nicotine. I wanted to fall into my bed and to forget about everything for a few hours. It wasn't too much to ask.

"Uh, no, they shot at my bodyguards. They were following in another car."

"And were they injured?"

"I don't think so. They ran off, and I haven't seen them since." I was wishing now that I knew their real names so I could warn the world against hiring them.

"But wasn't it your bodyguard who killed two of your attackers and forced the others to retreat."

"Yeah, but it was my other bodyguard, Ste—" I cut myself off when I saw him subtly shaking his head. "It was my other bodyguard. Like you said, he, uh, saved us."

"Do you know who they were?"

"I have no idea. I'm not sure why anyone would want to kidnap me— what they think they could accomplish by that."

"Maybe they think they could use you as a bargaining chip to get cigarettes back in the stores again," she said.

It was hard not to think about those guys mailing my body parts to Terra along with their demands. I could hear Paul Trainer now: *"Jesus Christ, Trevor, it's only an ear! You've got two on them. Quit being such a whiner."*

"I imagine they'd have been disappointed."

"How far away were you from your attackers when they were shot?"

"A few feet, I suppose."

"What was it like to see them killed so close to you?"

Honestly, I didn't know how I felt about the deaths of those men. Granted they were probably out to kill me, but still Stephen's casual efficiency had me a little confused. All this was so far from anything I'd ever experienced before.

"You have no comment," she prompted.

I shook my head.

"Do you think this is going to affect your discussions with the anti-smoking lobby and the government?

"As far as I know, there are no discussions," I said. "The government is ignoring the situation, and the antismoking lobby seems to have disappeared."

"But the economic ramifications of this shutdown are enormous. Surely the government will have to react."

I shrugged. "You would think . . . Look, I almost died a few hours ago, and I'm really tired. I've got to cut this short. Could y'all get out of my house?"

"Just one more question, Mr. Barnett. Do you have any information on the Ken Ewing situation?"

I knew the name, but it took me a few moments to retrieve it. "The manager of the manufacturing plant I spoke at? What do you mean?"

"You haven't heard? He was kidnapped right around the same time you were attacked."

I rubbed my eyes with my knuckles, wanting her out of my house more than ever now. "I don't know anything about it."

NICOTINE, UNAWARE OF everything that had happened, barked joyfully when I walked into the guest bedroom. The sound of it made me feel a little better.

"They had her shut up in here," Anne said. "But she's okay."

I dropped to my knees and held Nicotine's head, keeping her tongue out of striking distance and rubbing her ears. "I was thinking maybe you should stay here tonight, Anne. Stephen's going to be downstairs, so we'll be safe. Tomorrow we can figure out what to do about your security."

"Okay."

She was sitting on the edge of the bed in the ill-fitting jeans and flannel shirt the police had provided after confiscating her bloodstained clothing.

"So much for your breakthrough, huh, Trevor?"

I pressed my back against the wall and slid down to the floor. "Maybe it's time we rethink this, Anne."

"Rethink what?"

"This. Everything. What we're doing at Terra. I mean, it was a fun game for a while—to pretend we could change the world. By I'm not sure it's worth dying over."

"It wasn't a game to me," she said.

I propped my head in my hands and stared down at the floor. "Okay, what if it isn't a game? What if by some miracle we can change things for the better—save your proverbial one person? I'm not sure I'm willing to risk my life trying to help a bunch of people who can't help themselves. You know what? It's not my fault that people smoke. It's their own goddamn fault."

"So if it's not your fault, why don't you just quit?" There were any number of ways I could take that sentence, and her tone didn't offer any hints as to how she'd intended it.

"Who cares if you lose your trust?" she continued. "You're not making any money off it anyway, and at this rate you're not going to live long enough to get the distribution."

It wasn't a great job; that was for sure. Not only was I working for perhaps the deadliest corporation in the world, my primary responsibility was to draw fire that should have been aimed at Paul Trainer.

"What about you?" I said. "There's no reason for you to stay. You're not even getting paid."

"I'm not quite ready to quit just yet."

"Are you sure they're worth it?"

"Who?"

"Smokers."

She shrugged.

"I guess whether to stay or go is the easy question," I said.

"And the hard question?"

Nicotine slid her head into my lap, and I began rubbing her neck. "If the press was right and I really did have Paul Trainer wrapped around my little finger, what would we do with that power?"

A couple of minutes went by before Anne spoke again. But she didn't answer the question. "I thanked Stephen, but I haven't had a chance to thank you."

"For what?"

"For risking your life to save me."

I felt my face flush and found it impossible to meet her eye. I concentrated on petting Nicotine.

"So are we staying or going, Trevor?"

"Staying, I guess."

She dug a few loose pieces of paper from her purse and held them out to me. "Then it's time to start playing the game like we're trying to win."

"What's that?"

"The phone records from your father's office."

"How'd you get those?" I said, leaning farther back into the wall instead of reaching for them.

"It doesn't really matter, does it? Take them."

I did, albeit hesitantly, and focused on two highlighted numbers that I didn't recognize.

"The first call is to Angus Scalia a couple of hours before you went on TV with him. The second is a fax to his office."

After everything that had happened today, this wasn't what I needed.

"Trevor?"

"What?" I said, loud enough that even Nicotine gave me a sideways glance. "Is this supposed to mean something? My father could have been calling him to tell him he was going to sue him for slander or a thousand other things."

"He could have been, but he wasn't."

33 "So how are *you*?" Anne said.

Stephen stopped the car next to the elevator in Terra's underground parking lot and glanced down at himself as though he was looking for some hidden injury. "I'm very well, thank you. In fact, my wife and daughter are flying in this afternoon. They'll be here for the next few days, and then they're going to drive to the Grand Canyon. It's their first visit to America."

It was uncanny. Actually, *spooky* might be a better word. I'd initially dismissed the man's nonchalance as an act—a display of the Clint Eastwoodesque demeanor expected of a former commando. But it wasn't that. And it wasn't evil (an idea I'd flirted with briefly). He really felt no guilt or remorse over the men he'd killed. He'd done what he'd done and honestly didn't seem interested in soul-searching or second-guessing. I could learn a lot from this man.

"You're sure," Anne said and waited for his slightly perplexed nod before stepping from the car. I followed her without saying anything at all to him. The truth was that Stephen made me a little uncomfortable. Well, not him per se. More that this brutal assassin was one of the few people I'd met over the last few weeks who I kind of liked.

THERE WAS a lot of showy concern and back patting when we came through the glass doors of the executive floor. Secretaries stood behind their desks, executives wandered out of their offices. Even the coffee cart guy stopped and nodded respectfully. Neither of us reacted, other than to lower our heads and walk faster. When we arrived at my office, there was an enormous flower arrangement on Anne's desk with a card from Paul Trainer. I didn't ask what it said.

There were no flowers on my desk—just a note saying to join Trainer in his office ASAP.

I pushed through the door connecting our offices and saw that a meeting was already in progress. Trainer was talking quietly to a man I didn't recognize and seemed oblivious to my arrival. My father, on the other hand, strode across the floor and gave me a big bear hug. I flopped my hands around his back, but we were both aware that it was all just for show.

It seemed certain to me now that my father had called Daniel Alexander in Montana and told him to give me misinformation so that I'd crash and burn in front of the board. And when that hadn't had the desired effect, he'd faxed a copy of my trust to Angus Scalia in an effort to torpedo my first national television appearance.

I should have been furious. I should have stepped back and decked him, then blamed it on post-traumatic stress. But I really wasn't all that mad. If anything, I felt kind of sorry for him—which is probably the worst way you can feel about your father.

Parenthood had many noble traits, but one of the greatest was the hope that your child would surpass your successes. The jealously and competitiveness that my father felt toward me had always been sad, but now it seemed downright pathetic. Pathetic and dangerous.

"Are you all right, son? How are you feeling?"

"Real tired, Dad." We gratefully released each other. "But I'm okay."

Trainer was next. Finished with his conversation, he nearly ran across the office to frantically pump my hand.

"I can't tell you how relieved I am that you're okay, Trevor. I understand you showed incredible courage. Incredible! I really admire your fortitude in the face of all the death threats and lawsuits against you."

"What death threats and lawsuits?"

"Have you seen the press on this?" he said, ignoring my question and pointing to a stack of newspapers on his desk. "It's been excellent! Mark my words, son. We're going to win this thing."

He wrapped an arm around me, and we turned away from my father and the other man. "Now, what's with all that stuff you said about negotiating with the antitobacco people and cigarettes causing unpleasant deaths? I know that son of a bitch Ivers is tough and that we need to play up this benevolence angle for the public, but you're going to need to talk to me before you say anything like that in the future. No need to go off the deep end, here. Understood?"

He clearly wanted an answer, but I ducked out from beneath his arm and started flipping through the newspapers on his desk. There were a few obligatory articles implying that it would have been no great loss to the world if I'd been killed but overall, I'd come out of this thing looking pretty good. At this rate, it wouldn't be long before both I and the tobacco industry would soon achieve full antihero status.

"Trevor?" I heard Trainer say. I looked up slowly, waiting for him to try to get me to acknowledge his directive again. To my surprise, he didn't press the issue.

"I want to introduce you to Dr. Gregory Miller."

I crossed the floor and shook the man's hand, searching my mind for the familiar name. It came to me a moment later. Miller was a former CIA executive who did consulting work for Terracorp from time to time.

"It's nice to meet you, Trevor. I'm hearing a lot of good things."

He had dark circles beneath his eyes and a long, thin build magnified by jacket sleeves that were a little too short. Basically, he looked the way you'd expect someone with his background to look.

"Greg here is the one who recommended Stephen," Trainer explained.

"Thank you for that," I said. "Without Stephen, I wouldn't be standing here."

Paul Trainer pointed to the conversation pit at the edge of his office, and we all sat.

"What are we hearing from the police?" I asked no one in particular.

Miller answered. "They haven't been able to identify the dead men yet. No prints on record, and most of the phone tips the FBI's received are just crank calls."

"What about the two men who got away?"

"As near as anyone can tell, they just disappeared into thin air."

"Ken Ewing?"

"The police have even less on that—a couple of witnesses, but no physical evidence to speak of."

"So basically we don't know anything."

"I didn't say that," Miller said. "You asked me what we were hearing from the police."

Trainer spoke up. "We suspect that both attacks were planned and executed by Serbian terrorists. The cops'll figure it out eventually through Interpol, but they're not there yet."

Miller nodded. "The Serbs get a lot of their financing from cigarette smuggling, and you've cut them off from that income. We've been keeping our eye on them, as well as other organizations like Hamas who are heavily dependent on smuggling income . . ."

I kept my expression passive, but my mind was busy sifting through the information I was being given. Cigarette smuggling—essentially buying in places with low taxes on cigarettes and selling in places with high taxes—was a multibillion-dollar industry that quietly accounted for almost 10 percent of Terra's sales worldwide. It was also an activity that was subtly encouraged by tobacco industry management, who were more than happy to see their products made available at competitive prices all over the world. Undoubtedly it was our loose connection to these smugglers, and not Miller's brilliant detective work, that made it possible for us to be one step ahead of the authorities.

The attempt on my life and the kidnapping of Ken Ewing by foreign terrorists were giving Trainer an opportunity to rewrap the tobacco industry in the flag it had worn for so many years. Were the attacks just a circumstance that he was exploiting, or was it more than that? Was a decision about whether Ewing would be martyred or heroically rescued being made by our marketing department?

The fact that I had been saved (and made available for interviews a few hours later) didn't bode well for Ewing's future. The media needed new angles to keep things fresh and to keep people tuned in.

"What's going on with the Italians?" Trainer asked.

"The Mafia is by far the most sophisticated of the smuggling organizations, and that means they're smart enough to realize there's nothing they can do about this. They've got substantial cash reserves, and they're going to spend their money and energy maintaining their power base while this thing shakes out."

"What do the Serbs want?" I asked. "Do they think we're going to do an about-face because they're holding one of our plant managers?"

"I doubt it. More likely we'll get a ransom request."

"How much?" Trainer said.

"A lot. But I'm guessing they'll ask for cigarettes instead of cash—they need to keep their infrastructure in place and keep their competitors from trying to move in with Eastern Bloc or Asian stuff. Obviously, if this is the case, it'll make our lives easier."

It didn't seem obvious to me. "How so?"

Miller looked at me over his glasses, and my father followed suit. "Because it makes it possible to anticipate the mechanics of the exchange, Trevor. Cigarettes in the number they're going to want can't exactly be transported in a briefcase. The only way I can think of to work this would be to take a ship out into international waters and make the exchange there."

"Why do we care where it takes place?" I asked. "Shouldn't we be telling the FBI about all this? I mean, shouldn't they be here right now?"

"Relax, Trevor," Trainer said, leaning forward and patting my knee. "We're going about this in the most efficient way possible."

Miller stood and offered Trainer his hand. "If there isn't anything else, Paul, I should get back to work."

"Thanks, Greg. Stay in touch on this, okay?"

Miller disappeared through the door and before I could say anything, Xavier Rork, the director of marketing, came in. He didn't sit.

"How are we doin', Xavier?" Trainer asked.

"Things couldn't be better, Mr. Trainer." He nodded toward me. "A young, good-looking American man and his attractive assistant attacked in our heartland. We've leaked that we believe foreign terrorists are behind it, which obviously strikes a chord. The press is loving it."

"What about the tape?"

"Tape?" I said.

"Stephen had a video camera on his dash," Trainer explained. "Kind of like a police cruiser."

"There's . . . There's a tape?"

"Didn't I just say that?" He held a finger to his lips and pointed to Rork, who continued.

"We had our guys magnify and enhance it—the distance was fairly substantial . . ."

"Were we able to get a decent picture?" my father asked.

"Better than you'd think," Rork said, sounding a little unsure.

"And?" Trainer said.

"Frankly, we don't think it's appropriate."

"Why not?"

"Well . . . Let's just say your man was a little too effective. It plays like an execution. All our psychologists agree that it would backfire on us. You know, associate us with killing . . ."

Trainer nodded thoughtfully. "That's a damn shame. But, what'll you do, huh?"

"Give it to the police?" I suggested.

He grinned as if I'd made a joke that wasn't all that funny. "They'd just leak it, son."

"What about Stephen? I can tell you that the Montana police don't seem to like him."

Trainer waved a hand dismissively. "They don't have shit. And if they start making a lot of trouble, we'll suddenly find the tape. But they're not going to. There are four witnesses and enough physical evidence to fill a dump truck." He turned his attention back to Rork. "What about Ewing?"

"That's working out even better. He has a fantastic background: born to poor tobacco farmers, no college education, moving up from assembly-line worker to plant manager through sweat and elbow grease . . . He couldn't have been more appealing if he'd been handpicked."

I played that statement over again in my head: *He couldn't have been more appealing if he'd been handpicked.*

"We've put out a press packet on him, and his wife agreed to do interviews. Have you seen her on TV yet? I'll send you up a tape. She's a lovely woman, understandably distraught. We're moving forward with the angle that terrorists are trying to take control of this country but that we're not going to let them. The message is subtle, of course. It's easy to go over the top with something like that, so we're trying to keep it low-key."

"Did you see the front page of the *Times* today?"

Rork smiled widely. "That's the start of something. The *Times* isn't the only paper starting to focus on the fact that the government's actions regarding tobacco don't exactly jibe with what they say. Hypocrisy is a message we're going to keep hammering on."

Trainer nodded. "Keep it up, Xavier. You're the lead here. It's all up to you."

"Yes, sir," Rork said, performing a sort of military about-face and heading for the door. I watched him go, wondering what Ewing would think if he knew he was nothing more than a circus freak and his wife and kids were the hawkers.

Step right up now . . .

"And speaking of the government," Trainer said, "guess who called me this morning?"

He didn't wait for an answer. "That pissant, the White House chief of staff."

"What did he say?" my father asked.

"I don't know. I didn't take his call. I wanted the day's press to get out and saturate a little bit before I talk to him. Let him stew a little while."

34

I'D BEEN ON A WHITE HOUSE TOUR BACK WHEN I was in high school, and the only thing I could remember was feeling small and disconnected—like none of it had anything to do with me.

Things were different now. It was hard to feel anything but awe as we followed our pretty escort through the historic hallways. I was going to meet with the president of the United States—a man who had the same job as George Washington and Thomas Jefferson and Abraham Lincoln. The most powerful man in the world.

I carefully examined vases, rugs, light fixtures, paintings—even the elaborate molding—so that I could remember every detail. So that someday I could tell anyone who'd listen about my trip to meet the president.

"Did you vote for Anderson?" I whispered into Paul Trainer's ear.

"I don't vote," he boomed. "It just encourages the bastards."

BY THE TIME we stepped into the Oval Office, I'd spent the better part of an hour rehearsing my greeting. *"Nice to meet you, sir"* seemed too common. *"A pleasure to meet you, sir"* wasn't quite right, either, for some reason. I'd settled on *"An honor to meet you, Mr. President."*

I was surprised when our guide excused herself and left us standing in the middle of the room alone. Trainer stood with bored impatience, whereas I stood on my tiptoes trying to see what was on the desk. Attack plans for China? The latest data on a new terrorist threat? Information on Russian spies?

"Paul!" the president said, striding through a door to our left. "It's good to see you."

"Mr. President," Trainer said, calmly shaking the man's hand. "I'd like to introduce you to Trevor Barnett, my strategy guru."

I subtly wiped the sweat from my palm and shook the hand of the president of the United States. His grip wasn't overpowering and his hand was surprisingly small, but there was an incredible presence to him. Or maybe it was just the setting.

"It's nice to meet you, Trevor. I'm glad to see you're okay."

"Me too, sir," I fumbled.

He turned his back on me and put a hand on Trainer's shoulder, leading him to some chairs and a sofa set up in the middle of the office. I had the distinct impression that I'd already been forgotten. Clearly, Anderson had instantly seen through the diversion Trainer had constructed and instinctively knew what surprisingly few other people seemed to grasp. I was just a powerless figurehead. Whether that made him a brilliant man or just a brilliant politician was hard to say.

I took the least comfortable seat in the grouping because it seemed like I should. Anderson opted to just lean against a desk that I had read once belonged to JFK.

"Any word on Ken Ewing?"

"I guess the FBI is working the case, but they haven't told us anything yet," Trainer said. My heart rate rose a bit as I listened to him calmly withhold information from the president, and for a moment I wondered if there were devices that could measure that reaction.

"If there are any problems with the Bureau—if you feel like you aren't getting everything you need from them—I want you to come to me directly, Paul."

"Thank you, Mr. President."

There were a few seconds of silence before Anderson spoke again.

"How did we get here, Paul? This administration, and frankly every administration for the last fifteen years, has been nothing but supportive of the tobacco industry. And now you're running around threatening us."

"That's not true, sir."

I assumed that Trainer was referring to the implication that he was threatening. Stupid me.

"The federal government has been subtly chipping away at the tobacco industry for years—sometimes overtly by supporting suits and speaking out against us. Sometimes quietly by standing silent while we're attacked. In my mind, our relationship's become a little one-sided. You take billions of dollars from us every year and then turn around and treat us like criminals."

I held my breath, but no anger showed on Anderson's face. What Trainer had said was completely true but, as convoluted as it was, it was the natural order of things. His—our—recent actions had moved the government from their very comfortable position on the subject to one that was extremely delicate and more than a little dangerous. It was hard for Anderson to jump up and down in support of the return of cigarettes, but equally hard for him to ignore the tens of thousands of people out of work, the millions who wanted their cigarettes back, and the billions in potential lost revenue. Then there was the skyrocketing rate of petty crime and violence. I'd heard (but couldn't corroborate) that the producers of *Cops* were getting so much footage they'd had to hire college film students to keep up.

"What do you want, Paul? How do we make this go away?"

"I think you know the answer to that. We want legislation that will kill the lawsuits, we want a cap on taxes. Every time the government can't balance its checkbook or some politician starts feeling sanctimonious, we take the hit. A pack in New York costs over seven dollars, for God's sake."

"And how do you suggest I accomplish that, Paul? The Labeling Act is already in place."

What he was referring to was the Federal Cigarette Labeling and Advertising Act, which stated that warning labels had to appear on all cigarette packs, that they could be altered only by Congress, that they were sufficient, and that no one could sue arguing they weren't. Unfortunately, just about everyone found it convenient to ignore that law.

"Have you seen this?" Trainer said, pulling from his pocket a graphic photo of a human mouth rotting from cancer. "The Justice Department is suing to force us to add this picture to the government warning. So essentially the federal government is using civil action to coerce us to break federal law. Labeling and Advertising isn't cutting it, Mr. President. We want something ironclad."

"No one wants a showdown on this right now, Paul."

"No. No one does. But I'm about to lose a two-hundred-and-fifty-billion dollar suit that I'm going to have to settle. Where's it all going to end?"

"What do you want me to do, Paul? Completely revise the legal system? Threaten jurors? Make you above the law?"

"I want you to put meaningful legislation in place that will protect us from frivolous suits—both from the public and the government—and I want you to stand behind it. We agreed to settle with the states even though we could have won in the courts, because we were promised relief. And then the government just walked away from that promise."

"Before my time, Paul. I had nothing to do with that."

"That's the way of politics, though, isn't it Mr. President? The men in power do what's in their best interest at the time and the men down the line pay for it."

I was thankful to have been forgotten. While President Anderson was the picture of calm, and everything Trainer was saying was true, I wasn't sure it was wise to talk to the most powerful man in the world quite so directly.

"We're all victims of the political climate we live in, Paul."

"And the political climate wasn't right at the time of the settlement to offer us meaningful protection. I understand that. That's why we're working so hard to change the climate. In the past, smokers have tended to vote in lower numbers than nonsmokers. That's one of the things we're changing. I think that if this isn't resolved by Election Day, you're going

to see smokers turn out in droves. It should make for an interesting outcome."

Anderson blinked a few times. "Another threat, Paul?"

"Absolutely not, sir. I'm just telling you exactly what your campaign
manager probably told you two hours ago."

"Goddamnit, Paul!" Anderson said, the volume of his voice rising for
the first time. "You sit here and talk to me like Terra's selling tofu and
wheat germ. You've killed millions and lied through your teeth about it."

"Yes, sir. That's true. The question is what's the government going to
do about it?"

Anderson glanced over at me, but I looked away before I could see the
anger on his face.

"Why now, Paul? And don't tell me it's Montana. By the time that
precedent comes to roost, you and I will both be long gone. Why make all
this trouble now?"

Trainer leaned back in his chair, a concerned expression suddenly taking over his wrinkled face. "It's not about the money, Mr. President. It's
about the fact that despite every effort, no one seems to be able to figure
out that cigarettes are bad for them. We need to satisfy ourselves that
everyone understands the risks and accepts them in an informed way."

"Give me a break, Paul."

The more time I spent with Trainer the more I began to suspect that
this was about nothing more than proving to himself that he was still a
powerful, vital man. To make his mark on history before it was too late.

"This is a no-win situation for me, Paul. I've got plenty on my plate
without you playing these kinds of games."

"I disagree, Mr. President. I think that helping us is a no-lose situation
for you. Smokers want their cigarettes back and nonsmokers overwhelmingly believe that adults should be able to do what they want in their own
homes."

Anderson nodded toward me. "Your man here said that you'd be willing to make some concessions to get protection. What did you have in
mind?"

Trainer pursed his old lips for a moment, probably cursing me silently.

"We'll build an information campaign around the idea that this is a choice you make and there's no one to blame but yourself, and we'll work hard to get that message into schools. Hell, we're paying six hundred million a year on attorney fees—we can put at least some of our savings into getting the word out. I've got to think that it would be a more popular use of the money than paying a bunch of bloodsucking attorneys."

Anderson laughed. "And let me guess. That money would be deducted from what you pay attorneys, so if you continued to have high legal fees, the campaign would never materialize."

"I think that seems fair," Trainer said.

"And I assume you'd want *your* marketing department to design the campaign . . ."

"Obviously, we'd entertain input from outside people."

"The goal being to make it as ineffective as possible. Maybe ads where a bunch of stodgy old people scold kids through the TV. If you're clever enough you might even be able to design something that actually increases teen smoking."

Trainer shrugged. "We're all adults here. Of course, we're going to use this as an opportunity to sell our product—that's how we make money. And as you recall, that's how *you* make money, too. What's important is that we can make the antitobacco people think they won and make the public think you made a good deal."

Anderson nodded thoughtfully. "Put your people back to work, Paul. I agree with you—this is something that's going to have to be dealt with. But not by us. Let the next generation get bloodied in this fight."

"I can't do that, sir."

Anderson walked around his desk and sat. He obviously understood that Trainer wasn't going to back down.

"Okay, Paul. Fine. You win. You've proved your resolve and you've made your point. Send your people back to work, and I'll set up a task force to start putting together some legislation."

"Last time I heard that—and I realize it wasn't from you—I paid a quarter of a billion and got nothing in return. My people stay where they are."

Anderson let the calm façade he'd constructed crumble, and what was left was raw enough that it had me imagining government assassins sneaking up behind me with piano wire. I wanted to say something—to distance myself from Trainer—but I was too intimidated to speak.

"Goddamnit, Paul! You're going to send both your company and yourself down the toilet! For one second, could you just think about what you're doing and the people you're pissing off?"

Trainer stood and nodded respectfully toward Anderson. "I appreciate your time, Mr. President."

WHEN WE WALKED out of there, Trainer was grinning from ear to ear and almost shaking with excitement. He'd just strong-armed the leader of the free world. For a man too old for sex, drugs, and rock and roll, it was the ultimate rush. I, on the other hand, felt like I was going to throw up.

35 I WAS IN THE PASSENGER SEAT, BEING DRIVEN THROUGH my neighborhood by a guy who looked like an unsuccessful former boxer. Stephen was nowhere to be found, and I kept telling myself that he was spending time with his family and that this wasn't an indication that I was being set up (again?). I felt a little better when we rounded the corner and my house came into view, despite the three television vans still parked along the curb. I hung my arm out the window, using it to direct a flow of humid air across my face. I'm embarrassed to say that I still hadn't stopped sweating from my meeting with the president six hours ago.

"When are we going to get some peace and quiet!"

I glanced over and saw the old lady who lived four doors down from me standing in her yard shaking her fist at me. Seriously, shaking her fist. She'd been a complete pain in the ass ever since I'd moved in, once yelling at me because Nicotine barked through the door at her poodle every evening when she brought it to shit on my lawn. Even so, I'd always been unfailingly polite. Today I flipped her off.

"Straight into the garage," I said to my battered driver. A few reporters jogged up alongside the car, but they didn't seem all that fired up today. One had a cigarette in his mouth, and I wondered where he'd gotten it.

There were two messages on my machine. The first, from Lawrence Mann, said simply, "I heard you met with Anderson. Call me." The second was from Anne and was even less verbose. "So how'd it go with the prez?"

Nicotine bounded out of the living room and began rubbing against the legs of my dark slacks, leaving them covered with white hair. I knelt and put her in a playful headlock.

"You know what I did today, Nicky? I pissed off the president."

She broke free, as she always did, but instead of counterattacking she just sat down and cocked her head.

"That's right," I said. "He's probably organizing a death squad right now."

I dialed the number Mann left, listening to it ring as I headed for the kitchen to get something to eat.

"This is Mann."

"Larry. Trevor Barnett returning your call."

"How are you, Trevor? You're okay, right? And your assistant?"

Out of all the people who'd asked that question, he was the only who sounded like he cared about the answer. Sure, it was probably just a carefully constructed façade, but at least he had the decency to pretend.

"We're both fine, thanks. A little shaken up, but neither of us got hurt."

"I'm happy to hear that. So what happened with Anderson?"

"I don't think I'm supposed to be talking about that."

"Give it to me in broad strokes."

"I don't think anything was resolved."

"Trainer shot his mouth off, and Anderson got his back up."

"You said it; I didn't."

"Right."

"How are things going for you, Larry?"

"It could be worse. The banks are being friendly—tobacco is their bread and butter. That, combined with the aid the company's providing, means my people aren't suffering too badly. Not yet, anyway."

"Have you been able to catch the people pilfering smokes from our warehouses?"

He laughed. "No, but we're working on it. I'll keep you posted. Did you see the news?"

"Huh uh."

"That schoolteacher—the one from the Montana suit? She came out today in that wheelchair with the oxygen tank on it, and a bunch of people started booing her."

"You're kidding," I said, genuinely disgusted.

"I'm not. The mood of the country's getting downright weird. Even *I'm* avoiding going out in public. I'm not sure from one minute to the next if I'm going to get patted on the back or shot in it."

"Tell me about it."

"I met with Senator Randal today. He blew a bunch of smoke about how he'd laid the groundwork for your meeting with the president. That he's going to get us back to work."

"Truthfully, I think this has gone way beyond Randal. I wouldn't put a lot of faith in his ability to do anything right now."

"I'm not. I'm putting my faith in you."

I laughed. "That's even dumber."

"I don't think so. Move this thing, Trevor. Okay?"

My bodyguard suddenly appeared in the doorway to the kitchen and started motioning frantically for me to come with him. I ducked my head involuntarily. "Shit. Larry, I gotta go."

I ran after the painfully slow man, expecting to be headed for cover. Instead, he grabbed me by the arm and dragged me through the front door. We successfully negotiated the reporters on my lawn and ran toward a limousine parked by the curb. It started pulling away when I was still only halfway inside.

"You have to be having fun now!"

I'd landed in a less than dignified position on the floor and I pushed myself to my knees, then into the seat across from Trainer.

"Got something interesting to show you, son."

"Come on, Paul," I said, brushing dust and dog hair from my pants. "I just got home. Take me back."

"Too late! We're on the move!"

I glanced out the back window and then twisted around and looked past the driver out the front. We seemed to be at the center of a five-car motorcade.

"I think we found Ken Ewing," Trainer continued. "Miller was right. The bastards want cigarettes. They set up an exchange a couple of hundred miles offshore."

"Have you called the FBI?"

"No time," he said, grinning wildly. "We're going to have to handle this ourselves."

36 THE PRECISE MOMENT WHEN MY JOB DESCRIPTION changed from hapless industry spokesperson to hapless military operative, I wasn't sure.

Trainer's limo had taken us to the airport where we'd boarded the jet, then a helicopter, and now we were sitting in an enormous rubber raft with a nearly silent motor. The smooth glide of the craft suggested calm seas, but I couldn't visually confirm that. We were a few hundred miles off some unknown coast beneath a heavy layer of clouds that completely blocked out the moon and stars. It was an utterly perfect, completely balance-robbing, panic-inducing darkness that I'd never experienced before. I gripped a rope in front of me, though it was impossible to know if the other end was tied to anything, and tried to come to terms with my temporary blindness.

Paul Trainer was sitting next to me, and there were a number of other executives lined up on the wood benches behind us. Everyone was completely still and silent, heeding orders of the quietly dangerous-looking man who had helped us board and who I assumed was now at the controls of the outboard.

We sat there for a long time, occasionally getting splashed by waves breaking over the bow, but mostly just moving steadily forward through the blackness. Half an hour? An hour? Even time seemed distorted by the darkness. Eventually, though, I began to make out a distant glow that was so dim it disappeared if I looked directly at it. The quiet hum of the motor deepened a bit when we got close enough to pick out the profile of a medium-sized yacht.

A few minutes later, we pulled alongside, and the now-visible man piloting the raft lashed it to a ladder that he then quickly ascended. Two crewmen appeared a moment later and began helping us on board—no small task considering the average age and smoking habits of the other men in the raft. I have to say, though, that Trainer dragged his old bones up that ladder with the energy of a man on a temporary reprieve from death and determined to enjoy every minute of it.

There were five of us, not including our escorts, and none of us spoke as we were led along the barely lit deck and down a set of gloomy stairs. I had to shade my eyes when the door at the end of the corridor opened and we were ushered into a spacious room filled with soothing classical music and tables covered with elaborate hors d'oeuvres. There were no less than ten men already there, some familiar to me, others not. My father was grazing on some stuffed mushrooms in the corner.

"Paul! How was the trip?" Gregory Miller said, striding toward us to take Trainer's hand. "We can't get a helicopter in this close—it'd be seen. Was the raft all right? Fairly calm seas tonight."

"No problems, Greg," Trainer said. "Smooth sailing all the way."

"Excellent."

Miller stepped back and clapped to get everyone's attention. "Now that we're all here, I'd like to explain a little about our operation tonight." My father stabbed a few pieces of smoked salmon and got his vodka tonic freshened before rejoining the group. He didn't look at me.

"As most of you know, we received a ransom call from the men who kidnapped Ken Ewing, and it was more or less what we expected. They did ask for money, but primarily they were interested in product to keep their smuggling lines from collapsing. Also as predicted, they demanded that we steam a freighterful of cigarettes into international waters for a rendezvous tomorrow afternoon—at which time they plan to board it and sail it to an unknown port. With the knowledge of that rendezvous location and time, it was a small matter to locate the terrorists' vessel. It's currently about twenty-five miles from our position, bearing down on us at fifteen knots."

That got a quiet murmur going, and I looked back to see the worried expressions of the men behind me. When I faced front again, Miller was gesturing for calm. "No need to worry. They have no idea we're here, and they're not going to get anywhere near us. We have a team, led by a former SAS man, preparing to intercept."

My disappearing bodyguard, I assumed. Apparently, Stephen *wasn't* spending a few precious hours with his family; he was sitting in the middle of the ocean preparing to board a boatful of heavily armed Serbs.

"How?" someone behind me said.

"With the same type of raft you came in on. We've been able to get a schematic of the Serbs' boat and created an optimal plan to board and rescue Mr. Ewing. I'll go over that plan in a few minutes . . ." He motioned to a row of televisions bolted to the wall. "And we'll have a direct feed from the team's headset cameras, making it possible to watch the entire operation."

That got another murmur from the peanut gallery.

I WAS SURPRISED when no one stopped me from wandering off during the brief intermission meant to give Miller's aging audience time to avail themselves of the facilities and sample the chicken satays while they were still hot. I went up on deck, walking carefully along it and listening to the unfamiliar sounds of the sea. The only boat I'd ever been on had been a thirty-five-foot sailboat Darius owned during his very brief "chicks dig sailing" phase. He'd turned green and thrown up over the railing while we were still tied to the dock.

This was a little different. Instead of girls in bikinis, exotic drinks, and a nice solid plank leading to a nice, solid dock, all there was was silence and emptiness. I leaned out over the cold, slippery railing and tried to find something—anything—in the darkness.

When it became clear that there was nothing to find, I continued my circumnavigation of the boat in an effort to avoid going back down with Trainer and the others. I was almost back to where I'd begun when I came upon a small, utilitarian hatch in the deck. I confirmed that there was no one watching and then gave it a try. It opened easily, and there was enough of a glow coming from inside to see a ladder leading down.

I tried to convince myself to go back to the party but instead found myself climbing down the ladder and closing the hatch above my head.

The floor was about ten feet down, and when I got there I could see that what little light there was was coming from beneath a door at the end of a narrow corridor lined with pipes and electronics.

I couldn't really hear anything as I started forward but I could see motion in the light at my feet, suggesting someone was on the other side. It wasn't hard to guess who.

I was about halfway to the door when it finally occurred to me that I shouldn't be down there. I was about to turn around when I felt something round, cold, and metal press against the base of my skull. Sadly, the sensation of gun barrel on skin was becoming kind of familiar to me.

"Just stand very still," I heard the man behind me say.

Where he came from, I have no idea. There seemed to be only the one door, and I surely would have heard him climbing down the ladder behind me.

"I'm Trevor Barnett! I—"

Before I could get the rest of my explanation out, the door in front of me opened and partially blinded me.

"No, no! It's all right!"

I recognized Stephen's unlikely British lilt immediately.

The gun withdrew, but I still didn't move.

"Trevor. Nice of you to visit. Come in."

When I still didn't budge, Stephen's black-painted face broke into a bright white smile. "It's okay, Trevor. Come in."

None of the other men in the room seemed to notice when I entered, focusing instead on checking and rechecking the weapons lined up on the floor. Everything was much more compact than the enormous gun Stephen had saved me and Anne with. Tiny machine guns, grenades, knives, a crossbow . . . Seriously, a crossbow.

"I take it you're not going to the Grand Canyon," I said.

Stephen shook his head sadly. "I'm afraid I had to let them go on their own. Is there something we can do for you, Trevor?"

I didn't answer.

"Trevor?"

"They have hors d'oeurves," I heard myself say.

He laughed. "What?"

"They . . . They were bringing out those little quiches when I left."

His expression melted into one of understanding, causing the black stripes on his face to bend subtly. "It's not important."

"Yes it is. Doesn't it bother you that they're going to sit around and have cocktails while you kill people and maybe get killed yourself?"

He put an arm around me and began leading me back to the ladder. The man who'd snuck up behind me had disappeared again.

"Do you think it was any different when I did it for queen and country?"

"But what about your family?" I said. "What about your daughter?"

"They'll be well provided for if something happens to me. Paul Trainer has been very generous in that way."

"I didn't mean money."

I couldn't see his expression anymore—too much paint and not enough light. But he stopped at the base of the ladder and stood very still for a moment. "Maybe she'd find a father who was home more? One who's there to play tea party and help her with her homework? Maybe it wouldn't matter to her."

Back in college, whenever I hauled myself out of bed after a particularly impressive bender, I'd always ask myself the same thing: Am I crazy? Later, I started asking that question every morning no matter what—it seemed to be an important thing to monitor, though I'd never come up with a good answer. Now I thought I finally had one. Probably not, but everyone around me was completely certifiable.

The TV monitors were still dark, but everyone was staring at them anyway. Chairs had been set up and I was sitting on Paul Trainer's left, with Gregory Miller on his right. Miller continued to explain the operation—how Stephen and his men would board as stealthily as possible, how they would try to quickly and quietly incapacitate Ewing's captors so as not to endanger him. I tried to reconcile the word *incapacitate* with the equipment I'd seen Stephen's men sorting through, but couldn't quite lie to myself that energetically.

I thought about the Serbs' recent history of atrocities and what they might have done to me and Anne if Stephen hadn't intervened, but I still couldn't help feeling sorry for them. And I couldn't help feeling partially responsible for what was about to happen.

The monitors came to life as Stephen and his men scaled the side of the ship—a silent, shaky picture of mostly shadows and angles.

I'd love to paint what happened next in bold, heroic colors; to describe explosions and desperate hand-to-hand combat—to say that Stephen and his men made every effort to complete their mission without bloodshed. But that's not what happened.

Those narrow, quivering pictures showed men drinking coffee, playing cards, sleeping, shaving. Then there would be the elongated shadow of a black-painted blade and the slow grace of a lifeless body being gently lowered to the deck.

The monitor labeled STEPHEN reminded me of one of Darius's video games and I felt a familiar rolling sensation in my stomach as the camera moved down a claustrophobic staircase with the barrel of a gun bisecting the screen.

They found Ewing sitting on the floor in the center of a small room, his hands bound behind him. There was another man in the room, too, and he charged—only to be floored by a blow to the face from the butt of Stephen's gun. His predicament was made more permanent when Stephen separated his vertebrae with a practiced motion of his knife.

I was feeling pretty ill by the time Ewing was freed. I saw him stand, rub his wrists, and move his mouth in a silent thanks. I didn't watch the rest.

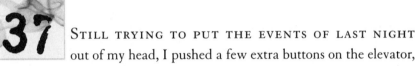

37 STILL TRYING TO PUT THE EVENTS OF LAST NIGHT out of my head, I pushed a few extra buttons on the elevator, causing it to stop randomly as it rose from Terra's underground parking garage. I stood passively at the back as the doors opened onto the silent, empty floors and then slid closed again.

When I finally got off, I saw that there were two new guards on either side of the thick glass protecting the executive reception area. They watched my approach, no doubt wondering about my untucked shirttail, wrinkled slacks, and missing tie, then used a punch pad to unlock the door.

"TREVOR, WHERE HAVE you been?" Anne whispered loudly after jumping up from her desk and grabbing hold of my arm. "Are you all right?"

I think she looked beautiful, but I can't remember for sure.

"I'm fine."

"You don't look fine. Where were you last night? I stayed up 'til midnight, but you never came home."

"You don't want to know."

Normally, there would have been no way she'd accept that answer, but on this particular day she just scrunched her eyebrows together and changed the subject. "Did you meet the president?"

I nodded.

She let me go, and I could feel her eyes on me as I walked into my office and closed the door behind me.

Three weeks ago I'd had a stable job, the illusion of friends, a position at Smokeless Youth that kept my soul from going completely black, and a woman that I futilely chased to keep my life from seeming like it just went around in circles. Everything had been going along so . . . evenly.

"Trevor! You watching this, boy?" Paul Trainer said, charging through the door. He dug through a drawer in my desk and came out with a remote control. A moment later a TV I didn't know was there began rising from an elaborate wooden trunk.

". . . there were bullets flying everywhere when I was evacuated." Ken Ewing looked a little more battered than I remembered. Both his eyes were blackened, there was a spiderweb of small cuts on the bridge of his nose, and his right arm was in a sling. "We had to detach the raft and pull away with two of our guys still on the boat, but we stayed about a hundred yards off the stern and waited. We saw them get blown over the starboard railing by an explosion about a minute later. Fortunately, we got them both out. One was—"

Trainer jabbed the Mute button when the picture changed to Ewing's joyful reunion with his wife and daughters.

I remembered the explosion, though not exactly as Ewing described it. I'd been standing silently next to my father watching the destruction of the Serbs' yacht cause a false dawn on the horizon. I'd looked over at him at one point and seen the reflection of the distant fire in his eyes, but he'd seemed to be somewhere else.

I assumed Ewing's account had been massaged by our marketing people and that the real story was that Stephen's team had set a bunch of explosives to send the Serbs, their boat, and any inconvenient evidence that might be on board straight to the bottom of the sea.

"That'll get your goddamn blood pumping, won't it?" Trainer bellowed.

"I guess so."

He gave me a disapproving frown and then hid his pain as he flopped a little too carelessly into a chair. "So? Not bad, huh? A heroic blow against terrorism and crime—courtesy of the American tobacco industry. How's that for turning lemons into lemonade?"

Did I really think Paul Trainer had set this whole thing up? I suppose not. But had he used the industry's connection with foreign smuggling to encourage it? Had he sent subtle signals through appropriate channels suggesting that kidnapping me and Ewing might bear fruit for the kidnappers? Had he endangered Ewing by withholding critical information from the authorities in hopes of a public-relations coup? Of course.

"I quit."

"What?" Trainer said, stiffening in his chair.

I know I told Anne that I'd stick it out, but last night's antics smelled faintly of cold-blooded, premeditated murder. I mean, realistically, what was the best she and I could hope for? To wield a nearly imperceptible bit of influence over how things were going to play out—influence that would quickly be identified and negated by Terra's marketing department? How far were we willing to go to act out what was, in the end, nothing but a futile gesture?

"I said I quit."

"You can't just quit. Not now."

"Yes I can. You know, I've never wanted to work here at all—I just did it because . . . because I never thought there was an alternative. And it was okay when I could bury my head in my filing project from nine to five and go home. But I never wanted to be part of anything here."

"Why the hell not?"

"Because in some ways, I believe the hype, Paul. I believe that we've knowingly sold death and then lied about it. I believe that we cultivate children as customers. I believe—hell, I *know*—that we've suppressed research showing links between smoking and disease."

I pulled my checkbook from my back pocket and flipped it open, showing him the crimson pen clipped inside. "I only sign checks in red, Paul—so every time I buy something I remember that it's with blood money.

And now I'm risking my life for . . . For what? Trust payments? To save the world from itself? To rescue the family business? To try to conjure up some self-esteem? It's not worth it."

Unexpectedly, Trainer broke into an uncontrolled bout of laughter that left him doubled over, coughing violently. I thought for a moment that I might be witnessing his death, but after about a minute he managed to regain control of his breathing.

"That's the most melodramatic pile of warm shit I've ever heard in my life. Christ, son, you can't quit. Without all this pain and guilt, who would you be?"

"I don't know," I said. "But maybe it's time for me to find out."

"Don't you ever get tired of being such a whiny putz?" he said. "It may surprise you to know that there are people in the world who actually have real problems and not just ones they make up every morning while they're shaving. What could you possibly want that you don't already have? Hell, it even looks like you're going to score with that girl from Smokeless Youth—and let me tell you that the office pool was giving hundred to one odds against you on that."

My eyes widened a bit, even though I tried to stop them.

"Yeah, I know all about her," Trainer said, standing and once again slinging a stiff arm around my shoulders. "You're being too hard on yourself, son. What if you burned every dollar you ever made? Hell, what if you donated it all to starving kids in China? What would it have changed?"

I opened my mouth to answer, but he cut me off.

"I'll tell you what it would have changed. Nothing. Except that you'd be poorer. People would still smoke. And they'd still die from it."

I squirmed uncomfortably, but he didn't remove his arm. A bone in his wrist was digging into my neck.

"I heard you on TV. You believe in freedom, just like I do." His arm tightened a little more, causing me to wince. "And now you're putting yourself at risk to stand up for that belief."

"You have no idea what I believe in," I said.

"I think I do. And I think you're getting to like the feeling of being off your knees and up on your feet. Why give that up?"

Trainer's assistant poked her head into my office. "Sir? I have Ken Ewing on the line."

"I've got to take this," he said, releasing me and starting for the door. "Why don't you just cool off for a couple of days. I don't think now's the right time for you to quit. I'm afraid you'd regret it later."

"I can quit anytime I want."

He stopped in the doorway. "I'll tell you what, Trevor. I'll make a deal with you. When I'm finished with Ewing, you come into my office and give me a coherent presentation on exactly what it is you expect to accomplish by quitting. If you can do that, I'll send you on your way with a nice severance package and a glowing letter of recommendation. But don't come in there and tell me that after years of working here and living off those red checks of yours that you're upset because you're actually involved. You were always involved."

He folded his arms across his chest and cocked his head, seeming to soften a bit. "You're no more damned today than you were a month ago. Trust me on this. I'm an expert."

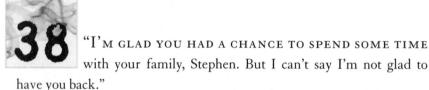**38** "I'm glad you had a chance to spend some time with your family, Stephen. But I can't say I'm not glad to have you back."

"I don't get to see them as much as I'd like," Stephen said as he swung the car into my neighborhood. "But now they're off to Arizona. Chelsea's at that age where large holes in the ground are far more interesting than her old father."

Anne was sitting in the passenger seat and I was in the back, second-guessing my decision not to take Trainer up on his invitation to let me argue my case for quitting. Anne and I had talked for a long time about the likelihood that Trainer had orchestrated the attack on us, but the conversation had been a little anticlimactic. In the end, neither of us saw much benefit to leaving now. It seemed a little optimistic to think that we could duck below everyone's radar so easily. Besides, Anne's one person was still out there waiting to be saved.

What I hadn't told her was what happened to Ewing's kidnappers. It wouldn't have changed her mind, and the truth was that I still felt kind of dirty from the whole thing. In the end, though, Trainer was right. My guilt or innocence had already been determined.

MY SEEMINGLY permanent media contingent had swelled a bit over the last twenty-four hours, and now the evening sun was glinting off no fewer than five satellite-dish-topped vans. The normal burst of activity that occurred when my car came into view was unnecessary today because the reporters were already in position, clutching their microphones and yelling at their cameramen. They must have posted a spotter somewhere.

"Not today, Stephen," I said.

I must have sounded odd because both he and Anne looked back at me. I saw him nod and felt the car accelerate.

Not only did we have the entire Ken Ewing rescue playing out all over the TV and newspapers, but a guy in Iowa also had taken his wife and kid hostage about four hours ago and was making no bones about the fact that if he didn't get some cigarettes quick he was going to blow them away. Last I heard there were something like fifty cops surrounding the place with snipers set up on the roofs of his neighbors' houses. Paul Trainer had graciously agreed to provide the smokes as long as the media had full access to the man's history of drug abuse, alcoholism, and violence.

By all reports, cigarettes were now pretty much unavailable in the United States. Supplies from initial hoarding had dwindled and the smuggling effort was fairly lackluster, since organized crime saw this as a temporary situation at best.

Of course none of this kept people from making an effort. In addition to our friend in Iowa, there were multiple reports of tobacco farmers resorting to salt-filled shotguns to chase away midnight harvesters, and travel agents were inundated with requests for extended vacations in countries where foreign-made cigarettes were still in abundant supply. All in all, things were getting pretty undignified.

I leaned away from the windows as the reporters clawed at them and shouted unintelligible questions. They chased us into the garage but when Stephen stepped out of the car, they all retreated back onto the driveway.

THERE WAS ONLY one message on my machine, and I unwisely pushed the button.

"Never hear from you anymore now that you're famous," Darius's voice boomed. "Having a party tonight. If you bring smokes, you'll be sure to score. Or maybe I'll have to take a hostage. How many cartons is that good for these days?" There was a long pause, and I imagined him starting to hang up and then changing his mind. "Oh, and if you come, bring that hot blonde from Channel Six. I think she's living on your lawn."

I noticed that Anne was glaring at the machine, but when she realized I was looking her expression turned distant. "A friend of yours?"

"I've known him since grade school," I said. "But I don't know if I'd call him a friend."

"You should go," she said with detachment that, believe it or not, sounded a little forced. "You deserve some time to unwind. And I'm sure an executive VP like yourself could scare up a few cigarettes somewhere."

There was a sudden shift in my reality that was disorienting enough to actually throw me physically off balance.

Anne was jealous.

There was undoubtedly something clever to say at this moment—something that would make it clear that I knew without being so overt as to put her on the defensive. Something encouraging.

"Uh, I'm really beat. I think I'm gonna just hit the sack."

Okay, that wasn't it. But it was the best I could do under the circumstances.

I CRAWLED INTO bed after a long shower and just lay there. The sun hadn't completely set and it was filtering through the shades behind my bed, creating stripes of light that penetrated my eyelids. I wanted to pass out from exhaustion, but it wasn't as easy as I thought it would be. There were too many things bouncing around in my head.

I spent half an hour like that before I got up, pulled on a pair of jeans, and padded out into the hall. I could hear Nicotine running around downstairs and Stephen encouraging her over the sound of the TV. Now it was my turn to feel a little pang of jealousy.

The guest-bedroom door was shut, and I stood quietly in front of it for a while. Now, you'd think that after lying in bed for half an hour, I'd have come up with that clever thing I should have said earlier, but I hadn't. Desperation and exhaustion seemed to be conspiring against me. I felt like I was exactly two inches from getting the only thing in life I really wanted but certain that I would lose it at the last moment.

I knocked, still unsure what I was going to say.

"Yeah? Come in."

She was sitting on top of the covers wearing bright green pajamas and holding a copy of *Time* with a lit cigarette on the cover. Some of her things had migrated from her apartment and were lying on the floor in milk cartons that last week had made up her living room bookcase.

"What's up, Trevor?"

I was a little hesitant but managed to get something out before it got weird. "I'm glad you're staying here."

She didn't smile or even put down her magazine. I was getting ready to back out and close the door again when she finally spoke.

"Me, too."

I'D FALLEN ASLEEP easily after that. The bed felt a little softer, the sun coming through the windows seemed warm instead of blinding, and Paul Trainer was a million miles away.

I WASN'T SURE why I woke up—the way I felt, I'd assumed that I wouldn't so much as twitch a finger until morning. It was sudden enough that I lifted my head for a moment but not so sudden that I felt the need to look around in the darkness. I fell back into the soft pillow, thinking about Anne, and started to drift again.

The reason I'd come awake the first time became a little more obvious when something poked me in the back.

"Go to sleep, Nicotine. We'll play tomorrow, I promise."

Another nudge. This one was hard enough to set me to bouncing on the mattress.

"Trevor!"

I rolled over and squinted at a slim, feminine figure next to my bed, then at my alarm clock, which read eleven-thirty.

"Anne? What's wrong?"

"I . . . I realized I forgot some stuff at my office at SY. We need to go get it."

I wasn't quite sure what she was driving at, and I flopped over on my back. "Get Stephen to take you over there at lunch tomorrow."

"I need to get it now."

"What?"

"I said, I need to get it now."

I examined her a little more closely and found that there was enough light to see that she was dressed.

"Why? What could you—"

"Just forget it!" she whispered loudly. "I thought maybe you'd help me, but I guess I was wrong."

That woke me up more or less fully, and I reached for the lamp on my nightstand. She grabbed my wrist before I could turn it on.

"What's going on, Anne? What's so important?"

"Will you help me or not?"

I sighed and threw the covers off, revealing an embarrassing pair of boxer shorts with Santas all over them before stepping unsteadily into a pair of jeans.

"Is Stephen still up?"

"Yeah."

"Go tell him we need a ride."

She just stood there.

I managed to find a pair of sandals and a golf shirt, and I put them on. When I started for the door, she grabbed my arm again.

"Maybe we could go out the other way," she said.

"Uh, there's only one door, Anne."

She quietly pulled the shade up and opened the window. I looked out into my swamp of a backyard but didn't move.

"Come on," she said. "There's a cab waiting for us a few blocks away."

39 "THIS IS GOOD," ANNE SAID.

The cabdriver eased over to the curb about a quarter of a mile from Smokeless Youth's headquarters. It was a little after midnight.

"How much stuff are we getting? Kind of far to carry anything heavy," I said. My grogginess was gone, replaced by mild nausea and a nagging headache that weren't so bad that I couldn't figure out it wasn't necessary to ditch Stephen and climb out a window to retrieve Anne's favorite pencil sharpener. So what were we doing here? I had no idea—Anne had ignored every question on the subject.

She handed the driver a wad of small bills and change that had the look of being the last of her savings. "Could you wait for us, please?"

"SERIOUSLY, ANNE," I said as we walked down the middle of the empty street. "What are we doing here?"

"You're just full of questions tonight, aren't you?"

I kept watch for nicotine zombies and terrorists as we continued on, deciding that I might as well just resign myself to going with the flow. I was still pretty good at that. Though maybe not as good as I used to be.

As we got closer, I could see that there were two cars parked in the otherwise-empty lot in front of SY's office building. One was an innocuous-looking Jetta that I recognized as John O'Byrne's. The other was an immaculate gray Cadillac the size of a small boat. I walked along the length of it, stopping again near the front windshield. I could feel Anne behind me as I stared down at a pack of Kool cigarettes on the dash.

Now, my father had never smoked a Kool in his life. However, he had an inexplicable paranoia that every black person in the country wanted nothing more out of life than to steal his Caddy. At some point years ago, he'd theorized that if he put a pack of Kools (apparently popular with the African-American crowd) on the dash, that the auto thieves hiding behind every rock would think the car belonged to a pimp and would leave it alone. I'm serious—this was his thought process.

Anne pressed her cheek against my shoulder. A moment later she slipped a hand in mine and started pulling.

This time I didn't budge. All I wanted was to go back to the cab, home, and bed. I was entitled.

She pulled harder, and I yanked my hand free. "I'm out of here."

"You can't leave!" she said, blocking my path.

"I can do anything I want."

"Anything Paul Trainer wants," she corrected.

I thumbed back to the Cadillac. "Why are you pushing this, Anne? You know what happens when you start sticking your nose into things like this? You find out a bunch of things that you didn't want to know and that you can't do anything about."

"How do you know there's nothing you can do about it? You have no idea what's going on in there."

I tried to get around her but she sidestepped, blocking me again. We did that little dance for a few more seconds before I gave up.

"Who do think has your best interests more at heart?" she said. "Me, your father, or Paul Trainer?"

I thought about that for a moment. "Is none of the above one of my options?"

She smirked and grabbed my hand again. "Oh, shut up."

We went around the back of the building and stopped at a window that was about five feet off the ground.

"What luck!" she said, feigning surprise as she pushed it open. "It's unlocked."

"What luck."

After giving her a leg up, I climbed in after her. There was a quiet thud as my feet hit the floor, and we both froze for a few seconds and looked around the dark room.

I saw the silhouette of Anne's hand as she put a finger to her lips but decided to ignore it.

"Come on, Anne," I whispered. "Let's just—" She clamped a hand over my mouth long enough for me to get the point and then walked over and stuck her head out the door into the hallway. I leaned over her and took a look for myself. It was empty and mostly dark. What little light there was was streaming from beneath the door of John O'Byrne's office.

Anne signaled for me to follow and began tiptoeing down the hall. When we got close enough to hear the unintelligible murmur of conversation, she stopped and turned back toward me. I held my hands up in a confused gesture. What were we going to do? Stick our ears against the door until someone opened it and we fell inside like an old Three Stooges episode? My vote was still to go home. Ignorance, in this case, was almost certainly bliss.

She carefully opened the door next to O'Byrne's office and went inside. I followed her into what turned out to be a neatly organized closet that went pitch-black when she closed the door behind us. I felt her hands on my shoulders and I let her pull me down on top of her as she lay down on the floor.

I have to say that, while a little weird, this was not an entirely uncomfortable position. I was holding most of my weight off her with my arms, but the entire length of my body was still pressing against her. You'd think I would have imagined us like this, but I never really had. It'd always seemed disrespectful. Even sacrilegious.

I was still trying to figure out exactly what was happening when she shoved a hand into my ribs and squirmed out from beneath me. There was a nearly inaudible scraping sound followed by a dim glow that partially illuminated the little storage room. I leaned in and saw that she had exposed a vent that led directly into O'Byrne's office.

My father's voice began floating through the opening, and I pressed in closer to Anne under the pretense of being able to hear better.

"You're overestimating the president, John. He's not any more outraged by the tobacco industry than he is the liquor industry. He's not on your side, he's not on our side—he's on the side of what's politically expedient. All he wants is not to get hurt here."

"But coming out strongly against the industry *is* politically expedient, isn't it, Edwin?" O'Byrne said, condescension and distaste clearly audible in his voice.

"Come on, John . . . Let's try to be realistic here. The antitobacco lobby's message right now is completely fragmented, and you've lost almost all your funding. Your support was pretty much universal as long as everyone was free to smoke and the money was flowing. But now . . . Well, attitudes are changing. You watch TV. You read the papers. This is all starting to move against you."

"I don't think—"

My father cut him off, as he had me so many times. "I've got to tell you, John . . . The president is pissed off right now, but he's not stupid. He knows damn well that if Trainer can hold out long enough, the government is going to have to cave."

There was a long silence as O'Byrne considered my father's perfectly reasonable analysis. Anne looked back at me, bringing her face to within an inch of mine. It's hard to describe how much I wanted to kiss her at that moment, but it didn't really seem like the time to climb out on that particular limb.

"There's an election coming up, and we've got a quarter of the population interested in only one thing—getting their cigarettes back," my father said. "And you can't turn on the television without having to listen to someone speculate that this thing is going to pull us into a recession if it goes on much longer. Nonsmokers don't want that, and most of them think smokers should get to make their own choices anyway. Wake up and look at the goddamn polls."

"I assume you have a point," O'Byrne said, sounding more than a little angry.

"I'm wondering what your best-case scenario is here? Is it that the industry goes under and no one can smoke anymore? That's not going to happen. And if it did—where would that leave you?"

As the victor, I thought. Anne leaned in closer to the vent, but when her old boss finally spoke, I don't think it was what she was hoping to hear.

"I assume you have something to propose?"

"I propose damage control. What you *don't* want is Trainer to win this thing outright. What if we could just make this all go away? Put things back the way they were? I've spoken to the president and he's behind me, but he wants support from the antitobacco lobby. He doesn't need a bunch of people going on TV saying he was too scared to fight and has doomed millions of people to die."

"What are we talking about?"

"We're talking about you saying that the industry was trying to put itself above the law, that he didn't blink, and that this is a win for America."

Another long silence.

"And what's in it for me if I agree to support this?"

There was no immediate response from my father, but I could picture him doing that smile/nod thing he favored when he perceived that he'd won.

"Increased financial support for the fight against underage smoking. We'll come to a firm number in the next couple of weeks, but I guarantee that you'll be happy with it."

O'Byrne didn't answer, despite Anne pressing her nose against the grate and willing him to.

"Come on, John. How likely is it that we're going to shrivel up and blow away? We're just too big and employ too many people. If Trainer wins and we get immunity from the courts, you'll lose the publicity you get from the big suits and everyone will start to see the fight against tobacco as a lost cause. You know it's true. Now can I tell the president that he has your support?"

Again no answer, but an agreement must have been reached because we heard someone stand.

"Good. You've made the right decision, John."

We heard a door open and then O'Byrne's voice.

"You haven't told me one thing, Edwin . . . What's in this for you?"

If my father answered, we couldn't hear it.

40 ANNE TOLD THE CABDRIVER TO TAKE US BACK TO her place and I didn't object, though in the back of my mind I wondered if we shouldn't be crawling back beneath Stephen's iron wing.

When we finally made it to Anne's apartment, I sank into the sofa and let my head fall back onto the cushions. She was clanking pans violently in her tiny kitchen, but I didn't pay attention. I was feeling a little self-absorbed.

It's a hard moment when you finally have to admit to yourself that your father is, beyond any shadow of a doubt, an ass. Not an evil genius or psychopath, not a bank robber or horse thief. Just a dime-a-dozen pathetic, mean, insecure prick. Honestly, if you find yourself in my position and have the capacity to lie to yourself on this subject, do it. Your life will be all the happier.

The undeniable truth was that he was working to undermine everything Trainer and I had been doing, and betraying the family business in the process. The answer to the question John O'Byrne had posed at the end of their meeting was crystal clear. What was in it for my father? Relevance.

I felt Anne tapping me on the knee and lifted my head. She handed me a cup of tea and I took it, even though I hate tea.

"I'm sorry," she said, sliding onto the sofa sideways and tucking her bare feet beneath my leg.

"For what?"

She shrugged. "When I started hearing rumors about that meeting, I tried to just ignore them. I decided I wasn't going to tell you about it . . ." She let out one of those short laughs that sounded like a precursor to tears. "I mean, I figured what was the point? But then . . ." Her voice trailed off.

"Don't worry about it, Anne. It's okay."

"Are you?"

"What?"

"Okay."

"Why wouldn't I be?"

I closed my eyes but could still feel her examining the side of my face.

"Maybe because your own father is working against you at the company? A month ago, I'd have believed you when you said you didn't care. But I know better now."

"My father isn't 'working against' me," I said, trying to convince myself more than her. "He's working against Paul Trainer. It's just business, Anne. It doesn't have anything to do with me."

For a moment I thought she was going to let me get away with that, but it wasn't to be.

"Fine. Sit there and lie to yourself if that's what you want to do. Sit there and let your father take over Terra and put you out on your ear. But you know as well as I do that a lot of this is about you. It shouldn't be, but it is."

The fact that she was right and that she wouldn't just let it go touched off something in me. I jumped to my feet and jabbed a finger in the air toward her.

"What the hell do you want from me, Anne? Do you want me to be like you? To give up everything I have to work for twenty grand a year at an organization that's whole purpose in the world is to make its employees feel superior? Do you want me to spend my life bringing coffee to a guy who just sold his soul to the general counsel of the largest tobacco company in the world for a few dollars and some TV face-time?"

I went on like that for long time. She just sat there wearing an expression of despair that I was too worked up to notice. When I finally finished my tirade, I fully expected her to attack me like she had that night at my house. But she didn't. For about thirty seconds, everything but the headlights washing through the window behind her was completely still. It was then that I realized I'd hurt her. It had never crossed my mind that I had the power.

"Anne, I—"

"I thought we'd gotten past this," she said, cutting off my apology. "I'm getting sick and tired of listening to you go on about what's wrong with the industry and the antitobacco lobby and the government and why it's not any of your business. I thought we'd decided to make it our business."

I flopped back down on the sofa, and to my surprise she slid her feet beneath my leg again.

It was hard not to wonder what she saw when she stared so hard at me. I assumed not much. I was becoming increasingly suspicious that there might be a reason I didn't have anyone I could truly call a friend. I'd told myself for the last twenty-five years that it was because everyone was shallow and blind and uncaring, but was that really it? Or was it because I was easy to write off? I'd spent most of my life trying to disappear. Had I been too successful?

"My mom didn't smoke before she had me," Anne said, breaking me from my trance.

"What?"

"She started in nineteen seventy. Kind of ironic, huh—just a few years after the government warnings came out. My father was a drunk—a man who makes your dad look like a saint. Mom would never say it, but I was an accident. I think it was the stress of trying to raise me and hold on to Dad that made her start."

"Anne, you can't—"

"She was always talking about quitting. Then my dad left, and we were on our own. It never seemed like she could put enough good years together back-to-back to stop. I can still see her sitting at our kitchen table, staring off into space with a cigarette in her hand. I think it was the only thing in her life that was ever easy—that gave her pleasure without asking for anything in return. But in the end, it did want something in return. On top of everything she'd been through in her life, she got cancer."

Anne leaned back a little farther into the sofa's worn arm. "For a long time I remembered her fighting, but now it seems like maybe it was me who was fighting. I'd skip school to take her to doctors, spend hours researching treatments, deal with medication, argue with insurance companies ... And she'd just sit there watching TV and sneaking cigarettes. Waiting to die. Maybe even looking forward to it, I don't know."

The headlights of a passing car beamed through the window and glinted off a tear that didn't quite have enough mass yet to make its way down her cheek.

I wanted to say something, but what? Should I start into my well-polished speech on freedom and self-determination? Go on about personal responsibility and blame? My many detractors were right. Those carefully crafted and unwaveringly logical arguments seemed perfectly relevant when you were talking about three hundred thousand people, but lost their potency when you were talking about one.

So I just sat there and listened. About her mother's aborted relationships, about lost jobs and lost hope, about constant moves and the struggle to fit in, about trailer parks and police. About how people die.

It finally dawned on me that Anne didn't blame herself for her mother's smoking or her death. Anne blamed herself for not loving her mother unconditionally. She blamed herself for the disdain and anger she'd felt and for the endless fights.

When she finished talking, I put a hand against her cheek and forced her to look at me. "Where do we go from here, Anne? You said a while ago that it was time to choose sides. Whose side are *you* on?"

She thought about it for a moment and then pulled herself onto my lap, straddling me and bringing her face down to mine. "I think we should talk about this later."

I PLAYED my game again that morning.

I lay there with my eyes closed, completely still and barely breathing. I didn't feel hungover but the memory of last night still had a fuzzy, dream-like quality. Had any of it really happened? Or would Nicotine run into my room in a few moments, jump on the bed, and prove for the hundredth time that I was alone.

I don't know how long I stayed there beneath blankets that felt too heavy and a sun that felt improperly angled, avoiding doing anything that might prove I was just lying in my own bed with an assassin in my kitchen and a pack of reporters on my lawn.

I finally mustered enough courage to slide a hand sideways and begin my search for resistance. It was a long two seconds, but my fingers finally contacted skin, sending a weak jolt of adrenaline through me. I ran them up a leg, a hip, a back, and finally slipped them into a tangle of soft hair.

I'd finally won.

Anne squirmed back and forth for a moment, trying to sink deeper into the mattress, and I opened my eyes. When I finally propped myself up on one elbow, she rolled on her side and peered up at me through eyes that were still half closed.

It was a moment that should have been ecstatic for me but turned out to be more a terrifying than anything else. What would I see when she finally came fully conscious? Regret? Horror? Disgust? Would she think I had taken advantage of her delicate emotional state? Had I?

But there was nothing like that. She just slid a hand across my stomach and let it lie there.

I know it sounds stupid, but I could have stayed in that bed forever. I wanted to talk more about her life, her mother, her plans. I wanted this morning to go on and on and on. But it couldn't.

I pushed the hair out of her face, and she smiled.

"You didn't answer my question last night."

Her head replaced her hand on my stomach, and she lay there thinking and waking up.

"Where do we go from here? I don't know. What I do know, though, is I want things to change. I want to end the inertia. I want to do something that makes a difference . . . I'm honestly not sure I care if it's a change for the better or for the worse. I mean, maybe you're right. Maybe if people knew that smoking was *their* decision and their decision alone, it might jump-start things again. Breathe some life back into us."

"Maybe," I said.

"At least we'd have a fresh angle to work from. We could try a really no-nonsense campaign to promote personal responsibility—"

"You're on your own, baby," I said, paraphrasing an old Virginia Slims ads.

She rolled on her back and looked up at me. "Hey, that's not bad. Can I use it?"

"Be my guest. So what if we take Paul Trainer's side? Would you be willing to risk Tobacco getting everything they ever wanted just to break you out of your rut?"

"I was hoping you could work it out so the industry has to make huge concessions."

I laughed. "Not likely."

It took a few minutes, but she finally gave a short, slightly defeated nod. I reached for the phone and dialed Paul Trainer's home number. It was Saturday, I remembered, and I figured I'd try him there first.

"Hello?"

"Paul? It's Trevor."

"Trevor. What's going on?" He sounded a little subdued, but I didn't think anything of it at the time.

"I need to talk to you."

"About what?"

"About my father."

"Talk, then."

"Not over the phone. Can I come over?"

"Sure, why not? I'll be here."

41 WE FOUND PAUL TRAINER ON THE WEST SIDE OF A house big enough to use the points of the compass to describe it. He was in a room full of flowering plants, tending them with awkward and uncaring movements that should have told me the setting was staged. I guess I was too excited by my newfound sense of purpose to notice the obvious clues. What I *did* notice, though, was that, surrounded by the expansive grandeur of his home, Trainer looked kind of small. A little more like the fading old man he probably was.

"Come in, boy," he said, waving at me with a dangerous-looking pair of shears. "What's so important that you have to come over here on a weekend?"

Anne stayed close behind me as I stepped through the door and into the floral-scented room. She swore that she was just tagging along to offer moral support, but it was more likely that her purpose was actually to make sure I followed through. Either way, it was good to have her strength to draw from.

"I needed to talk to you," I fumbled.

He picked a bug off a delicate fern and accidentally took the leaf with it. I felt a sneeze coming on.

"I think we established that. You said it was about your father?"

I nodded, though he wasn't looking at me. Anne whispered, "You're doing good. Go on," in my ear.

"He's planning to take over Terra and put cigarettes back on the market."

"Is that right?"

I nodded again.

"And what makes you think that?"

"I . . . *we* heard him meeting with John O'Bryne at Smokeless Youth. He's trying to get the support of the antitobacco lobby so that they'll back the move and stand behind the president."

He went back to his plant and didn't say anything. After about a minute, I looked over at Anne. She just shrugged.

"So what do you think, Paul? What should we do?"

"There's nothing you can do."

Oddly, I didn't react to the voice except to turn toward it. My father was standing in an open doorway to our right, leaning against the jamb.

"You're too late, Trevor—I've got everyone in my hip pocket. I warned you about playing a game you weren't up to."

Anne squeezed my shoulder as I looked back at Trainer. He was concentrating on his fern.

"That's right, Trevor. He called me," my father said, bouncing off the jamb and swaggering toward me. "He told me you were coming."

It occurred to me that Dad didn't have to be here—that he'd gone out of his way to create this face-to-face confrontation. The quiet drama of it was obviously calculated to magnify the humiliation of my defeat, but it had the opposite effect. I suddenly felt so completely superior to him.

"This wasn't personal, Dad. It's never *been* personal. I thought continuing with this thing was right—for the company and for the country. That's all."

"What you thought wasn't really relevant, though. Was it, Trevor? Don't you understand? You were never a player in any of this. Trainer used you and now he's throwing you away." He laughed humorlessly. "You've turned out to be quite the hypocrite. You spend your entire life acting like you're too good for all this, but when you got a taste of the power you'd never been strong enough to go out and take, you lapped it up. Isn't that right, Paul?"

Trainer nodded as though there was someone controlling his head with a string, and Anne leaned into my ear. "I think maybe it's time to go, Trevor."

But she was wrong. It wasn't time to go. I felt ashamed that I'd never really stood up to my father—that I'd let this little man play out his insecurities on my mother and me unchallenged for so long.

"Why did you send Scalia that copy of my trust, then, Dad? I mean, if I'm so irrelevant, why bother? Were you trying to prove to yourself that you're smarter than me? Stronger than me? You're not. You never have been."

His expression hit me hard. I'd swear I saw hatred in it.

"No? Do you have any idea how easy it was to outsmart you? I've been talking to Randal and the other senators ever since Trainer blindsided them by shutting down. And then you walk into the oval office and insult the president? Brilliant move. By the time you got on TV and started going on about concessions the board was calling *me*—not the other way around. If—"

"What's your problem, Dad? I've never understood it. Is it that Grandpa liked me and your brothers better than he liked you? News flash, Dad: Grandpa was a prick and had the depth of a mud puddle. You're nearly sixty. This 'my daddy didn't love me' thing is starting to look a little stupid."

"You're out, Trevor!" he said in a voice now shaking with rage. "Don't come into the office on Monday. Don't ever let me catch you in that building again."

He turned and started for the door, but I ran over and blocked his path. The final, irretrievable loss of my job and trust had less of an effect on me than I thought it would. What really pissed me off wasn't my newfound poverty; it was the fact that my trust would revert to him.

"Didn't Grandpa leave you enough money, Dad? Now you need to take mine, too? I guess at least you can go to sleep now knowing you're richer than me." I paused. "Oh yeah, but you didn't make any of that money, did you? Someone gave it to you."

"I worked my ass off to get where I am!" he shouted. "I'm the god-damn CEO of Terracorp!"

"And I was the executive vice president of strategy. Big deal."

"Well, we're about to see how far that gets you, aren't we?" He turned that weird, washed-out purple he always did when he was really furious, but wouldn't meet my eye. He turned and pointed at Anne. "And don't you bother trying to crawl back to O'Byrne. I'll be on the phone to him in ten minutes."

Her nose scrunched up in an expression that you might get if you accidentally stepped on a slug barefoot. "I guess I'll have to go back to making a quarter of a million a year suing you, then."

When he went for the door again, I stepped aside and focused on Trainer again. "I take it you're not going to fight this."

"We had a good run, boy. But I can't afford to fight."

"How much are you getting?"

"Enough."

"What about me?" I said.

He grabbed a watering can and began drowning an orchid. "What about you?"

STEPHEN, WHO'D DRIVEN us there, was nowhere to be found when we stepped back around the front of the house. He'd undoubtedly abandoned us to protect my father's much more important body, but had the good manners to leave the keys in my ignition. Anne and I took a circuitous route home, using well-traveled roads that I wouldn't normally take. The announcement that the cigarette floodgate would soon open again hadn't yet been made, and it seemed likely that I still had a lot of enemies.

By tomorrow, though, I figured every smoker in America would have a cigarette perched between his or her lips and the media would move on to a good political sex scandal, missing child, or immigrant who'd seen Jesus' face in a tortilla. And honestly, that was okay with me. Having taken my best shot, I was at peace with the idea of fading into the background again.

"That could have gone a little better," Anne said, testing the waters. She clearly felt a little responsible for the showdown with my dad, but the truth was that it had been long overdue.

"Your father's a businessman," she said. "That whole game is about getting ahead . . ."

I smiled, surprising myself with how easy it was.

"What about your trust?" she asked, sounding increasingly despondent about her role in maneuvering me into my current predicament. "Can he really fire you?"

I didn't know. As a practical matter, it seemed like I'd have substantial grounds for a wrongful-termination suit. I mean, I'd risked my life for that company and done everything asked of me. If those weren't the actions of a pretty damn good employee, what were?

The problem was, I had a feeling that that was exactly what he wanted. He wanted to see me grovel, and once he did, he'd be happy to enslave me again by giving me the most demeaning job he could find.

"You weren't just gold-digging, were you?" I said. "You're not going to dump me now that I'm poor."

Her smile looked more relieved than happy. "I probably would, but I sense good income potential."

That actually made me laugh. "I'd hate to meet someone you thought had bad income potential."

"I'm serious, Trevor. Think about it. You're the thirty-two-year-old former executive VP of one of the largest corporations in the world. You're all over TV and the newspapers. Don't sell yourself short."

"I admire your optimism."

"Monica Lewinsky's a millionaire."

"Thanks for that flattering analogy."

"My point is, there's no such thing as bad publicity."

She had a point, but I wasn't sure that was what I wanted.

"What about you, Anne? Are you serious about going back to being a lawyer and suing the tobacco companies?"

She settled into her seat and watched the landscape speed by through the passenger window, but didn't answer me.

42 "I CAN'T BELIEVE I LET YOU TALK ME INTO THIS."

Anne rolled her eyes and kept moving forward through the soggy grass.

Except for the reverse angle, everything was as I remembered it. The elaborate podium I'd stood on a week earlier was still there, framed by the enormous cigarette manufacturing plant and elevated to a majestic height by the natural slope of the land. The audience, though, was at least twice the size of the one that showed up to hear me talk. I didn't take it personally.

"Don't you think this is far enough?" I said, using my elbows and shoulders to keep us from getting bogged down in the crowd.

"Quit being such a baby," Anne said. "Don't you want a good view?"

"No."

She ignored me and continued her surgical penetration of the sweating mass. In my effort not to make eye contact with anyone, I tripped and nearly did a nosedive into a little kid. By the time I'd gracelessly recovered my balance, the people around me were all staring. I froze and waited for something to happen, but nothing did. A moment later, I felt Anne's inevitable tug on my hand and moved on.

Word of my presence spread faster than we could walk and soon the crowd was parting in front of us, creating a corridor just wide enough to pass through. A man in a pair of worn overalls pulled his daughter behind him as though he thought I might attack. A woman with horn-rimmed glasses folded her arms and glared at me. A kid who looked like a high-school football player stepped forward to prove he wasn't afraid, then retreated.

It took another five uncomfortable minutes for Anne to position us dead in front of the lectern. We were still about twenty feet away, but that was as close as the yellow barricades and guards would allow. I glanced at my watch and saw that, as I had been, the main attraction was late. We were dying in the sun and the sweat was starting to soak through my clothing, leaving less than dignified stains growing on the thin fabric of my shirt. Anne had had the presence of mind to wear an enormous hat that shaded most of her body and made her practically invisible from above. It was dark purple and, combined with a similarly colored sundress, it made her look like a bridesmaid at the wedding of a grape. I, of course, kept this to myself.

Another five minutes went by before my father, all youthful energy and perfect hair, finally bounded up the steps of the podium. The air filled with a bland round of applause with one noisy dissenter. Anne began booing loudly.

I nudged her. "Shut up. You're going to get us in trouble."

She tilted her head back, and her face appeared from beneath her hat. "Try it. It's fun. "

I had to admit that the idea was kind of appealing. I cupped my hands around my mouth and gave a test boo. It felt good, so I threw a little more feeling into the second one. My father was intently oblivious, but everyone around us took another step back.

"Hello, I'm Edwin Barnett. Many of you already know that I've been named to replace Paul Trainer as the new chief executive officer of Terracorp."

The clapping died off, and I put a hand over Anne's mouth. She bit me.

"I know these past few weeks have been hard on everyone and on behalf of the company, I want to apologize for that. My first order of business as CEO is to get everyone back to work."

That got another round of applause, but not the rousing thunder you'd expect.

"I'll be honest with you . . ."

Seemed kind of unlikely.

". . . there's been some serious infighting going on at headquarters about the direction this industry is going to take. And I'm afraid that y'all were the victims of that. Be aware, though, I'm sensitive to how hard this has been."

As an industry spokesman, it occurred to me that I'd been underpaid. My father oozed insincerity as only a tobacco executive or used-car salesman could.

"I've worked closely with the other CEOs and the government—most notably President Anderson and Senator Randal—to put an end to this stalemate." He paused dramatically. "As of tomorrow morning, we're open for business again."

Another lackluster round of applause. I looked around me at the worried faces turned up toward my father.

"I want you to know that the sacrifices you made over the past few weeks weren't wasted. We made our point. I have a number of meetings scheduled with President Anderson to start talking about how we can make our companies and our jobs more secure."

"What exactly does that mean?" Anne shouted, loud enough to silence the few individual hand claps that were threatening to spread. "It sounds like you folded like a cheap suit! Did you give yourself a big, fat raise when you took over and sold us out?"

In a testament to the efficiency of the tobacco industry, no fewer than four bulky security guards suddenly materialized and grabbed us. I didn't bother to struggle, but Anne thrashed around like a hooked trout. "Don't believe him!" she shouted, apparently having a hell of a good time being dragged through the crowd. Trainer was right. I really needed to learn to quit worrying all the time and enjoy life.

"You think he cares about whether you starve or not? He just—"

One of the guards put a hand over her mouth and she bit him, too, but

I suspect a lot harder. As he drew a hand back to hit her, I elbowed the guy holding me and sent him staggering backward. I was gearing up to swing at the guy about to hit Anne when Stephen suddenly emerged from the gawking crowd and the guards immediately retreated. He put a hand on each of our backs and pushed us toward the parking lot as my father's amplified voice started in again.

"I'm sorry about that. We tried to keep the antismoking lobby out of here, but we can't catch 'em all . . . As I was saying, I'm having meetings with the president where . . ."

"Who says being a sore loser isn't fun?" Anne said when we finally cleared the crowd.

"Well, I think you've both had quite enough fun for one day," Stephen said, continuing to push us gently toward my car.

She looked back and gave him a guilty frown that was entirely too attractive. "Sorry, Stephen. Are you going to get in trouble for not giving us the rubber-hose treatment?"

"Who said I wasn't?" He sounded serious enough to worry us both but then broke into a smile. "I think maybe a strategic retreat and some regrouping might be in order for you two."

He finally stopped at the edge of the parking lot and gave us one last shove. "Good luck."

"See ya, Stephen," Anne said.

I picked up the pace in case he changed his mind about the hose.

"Where were you back there?" Anne said, jabbing me playfully in the ribs.

"Right behind you."

"Kinda quiet."

"I booed!"

"You call that a boo? Let me hear it again."

"Boo," I said in a quiet, irritated voice.

"Come on!" she said, turning sideways and skipping along next to me. "Boooo!"

"Boooo!"

"There you go. I'll make a tobacco industry heckler out of you yet."

We climbed into my car and I pulled out, carefully maneuvering through the tightly packed cars on our way back out to the road. We'd only made it about twenty feet when I heard the back door open and someone jump in. I jammed on the brakes, and the man pitched forward.

"Ow! Is that any way to treat an old friend?"

I wasn't sure what to do, so I just started driving again. Our passenger stuck his hand up between the seats and offered it to Anne. "Larry Mann—I work for the union."

"Anne Kimball . . . I'm unemployed."

He leaned back as I eased out onto the rural highway. In my rearview mirror I could see a carful of what I assumed were his men following us.

"So what happened, Trevor?"

"What do you mean?"

"I mean with you and Trainer."

I shrugged. "Apparently there was a sudden change in management philosophy."

"I gathered."

"It seems that the thought of you guys going hungry was keeping my father up at night. He just couldn't bear the strain."

"Sarcasm," Mann said. "It's new on you, but you wear it well. Actually, I've talked to your father."

"Really? What did he say?"

"Oh, he told me how important we are to the company, how he's looking out for us, how much he respects me . . . You know the drill."

"Yeah, I suppose I do."

"I've got to tell you, Trevor, that man is as big a prick as I ever met. I sincerely hope your mother was cheating on him when you were conceived."

"Thanks," I said.

"So what now?"

I looked over at Anne. "Disneyland?"

had a real opportunity for everyone in this country to get on the same page where the risks of smoking are concerned. Unfortunately, when push came to shove, the powers that be didn't share my vision."

"Do you feel that this is a step backward for the interests of the tobacco companies?" I recognized the guy who'd asked that question. A financial reporter—I couldn't remember from where. It was hard not to think about the fact that what I said next would either make or lose millions of dollars for tobacco industry shareholders—a group I was no longer a member of.

"Honestly, I don't think it's in the best interests of anyone, but it's the board's call and I have to respect their decision."

"Would you elaborate on that, Mr. Barnett?"

The competition in the group seemed to be easing as it became clear that I would keep my word and answer all their questions. The increased calm allowed me to relax a bit.

"People know that smoking is bad for them—if they can't tell by the way it makes them feel, then they can read the big warning on the pack. What we've lost here is an opportunity to create an environment of personal responsibility that perhaps could have reduced the smoking rates in this country, stabilized an economically critical American industry, and stopped wasting the courts' time in endless lawsuits."

"Or an opportunity to completely do away with the tobacco industry," someone said. I saw Anne slip out of the car and make a break for the house. No one else noticed.

"Maybe," I said. "That was for the American people and their representatives to decide."

I felt like I could now check off "unflappable."

"But isn't it true that your real goal was to create an environment that would allow the tobacco companies to sell a dangerous product without fear of litigation?"

"In a word—yes. It's true that we make a dangerous product. If you smoke, there's a pretty good chance it's going to kill you one day. And if

Mann laughed. "You're turning into a real tough guy, you know that? What happened to the confused little boy I met a few weeks ago? I'm serious, Trevor. You're the only guy in this whole goddamn industry who even understands the concept of a straight answer. Give."

"Turns out Paul Trainer's nuts," I said.

"And this is news? What I want to know is how all this fell apart. You'd met with the president, you had Congress worried about their jobs, the press and the country were coming over to your side. Why give up now?"

"My father is afraid of becoming irrelevant," I said. "And the board's scared."

"I see . . . Well, what about us, Trevor? What happens to us?"

"You got what you wanted. You got your jobs back."

"But for how long? We're going to lose in Montana. How long until the industry gets sued into oblivion? How long before Terra can't raise capital? Can't get loans? When we start to get squeezed, it isn't going to be management that feels it; it's going to be my people."

I shrugged at him in the rearview mirror. "I feel for you, Larry, but none of this is my problem anymore."

43 "WHAT'S UP WITH THIS?" I SAID, BANGING ON THE steering wheel in frustration. Instead of dwindling, my media contingent had doubled in size, and the appearance of my car had reporters jumping from their vans and scurrying around as though someone had thrown a grenade.

"Look at my lawn, Anne! How am I gonna sell this place now that they've destroyed my damn lawn?"

"Maybe you could put in one of those Japanese rock gardens."

I slammed on the brakes, bringing the car to a lurching halt a hundred yards from my driveway. "What the hell's wrong with them? I'm fired and cigarettes are going back on sale tomorrow. I don't have anything else to say." I shoved the car into reverse and was about to do an impression of my former bodyguards when Anne pulled the shifter into neutral.

"Hey!" I said over the sound of the revving engine. "What are you doing?"

"Maybe you should talk to them."

"About what? I'm not the company's spokesman anymore. I'm just a tobacco industry heckler, right? And not even a good one."

"You've got potential," she said through a broad smile. "It just takes a little practice."

Through the windshield I confirmed that the reporters weren't advancing on us. Yet.

"Have you ever thought of just speaking for yourself, Trevor? It might be a good time for you to get on TV and sound magnanimous and intelligent and unflappable. To tell people what you were trying to do."

"What's the point?" I said.

"First of all, I'll bet it will make you feel better. And even if it doesn't, you never know when a future employer might be watching. It seems to me that the combined income of the people in this car is zero. And I like to eat. Remember?"

Combined. I liked the sound of that.

I put the car in drive again and glided regally forward, trying to get in a magnanimous, intelligent, and unflappable frame of mind. It turned out to be kind of a stretch.

The reporters crowded the car, shouting and attacking the windows with their microphones. Instead of pulling through them into the garage, like I normally did, I forced my way out onto the driveway. I motioned for quiet and, for some reason, it worked.

"Okay," I said loudly. "I'm going to stand here and answer every question you've got as honestly and completely as I can. But after that you've got to leave and let my lawn heal, okay?"

Very magnanimous.

"Mr. Barnett—" one of them started, but I held up my hand for silence again. "Let me just say a few things before we start the Q and A. First, I am no longer employed by Terracorp and so I no longer speak for them."

"Were you fired?"

I thought about that for a moment but couldn't see a way to sugar coat it. If anyone in history had ever been fired, it was me. "Yeah, I think it would be fair to say that I was fired—"

"By your own father? How did that make you feel?"

"Unemployed. It made me feel very unemployed. Look, the strategy that Paul Trainer and I came up with"—I figured I might as well take some credit since I was in the market for a new job—"turned out not to be viable in the board's eyes and they've decided to go back to the company's former policies. Frankly, I don't agree with that decision. I thought we that bothers you—if you don't think it's worth it—don't take up the habit. If, on the other hand, you feel it improves the quality of your life to the point where you don't mind it being potentially shortened, smoke up. But I'll tell you that we weren't looking to get something for nothing. We were willing to bring in the antitobacco lobby and the healthcare community to help create a compromise plan that would hopefully take the issue of public health out of the courts and put it back on the legislative side of the government where it belongs."

No one would be able to disprove that statement, though of course I knew that Paul Trainer wouldn't have given an inch to the antitobacco lobby. Intelligent-sounding, though. I could almost feel my answering machine filling with job offers.

"Then what's your reaction to the strike?" the blond Darius was so infatuated with said.

"What?"

"The strike?" she said, enunciating a little more carefully this time.

"What strike?"

They all looked at one another. "You aren't aware that Lawrence Mann has called a strike by the Tobacco Workers? They started picketing production and distribution facilities about half an hour ago."

I started to lose my grip on the unflappable part of my façade as my mind tried to wrap itself around that new piece of information. I cleared my throat as the woman, sensing weakness, edged toward me with her microphone stretched out in front of her like a weapon. "Apparently, the tobacco industry's food divisions have also walked off the job."

"My theories on doughnuts must have had them worried," I managed to get out. Everyone laughed.

"Well, you obviously figured out a way to scare the shipping industry, too, because the Teamsters are honoring the strike by refusing to deliver tobacco products."

"I, uh . . . I told you I'd answer your questions, but I'm not sure what to say here. I didn't know anything about this."

The blond woman motioned toward her van, and we all followed her. The back was open and she leaned in, speaking to the guy sitting inside. "Bobby, can we get a playback on Lawrence Mann's statement?"

A moment later, Mann's face appeared on a tiny four-inch screen. I watched it while everyone else watched me.

"We supported Paul Trainer and Trevor Barnett in this from the beginning, and frankly we think it's important to finish what we start. The tobacco industry—our way of life—is being slowly destroyed. But with no real goal other than to make a bunch of attorneys millionaires. We work hard down here, and we sure as heck aren't rich. But we do have families and we have to think about their future. If America wants cigarettes called a narcotic and made illegal, then it's time to stand up and say that. On the other hand, if we want to continue with freedom and the right to choose, then we need to say that. But it's time that the people of the South know if we're going to be able to feed our kids and provide for ourselves in our old age. And that decision shouldn't lie with twelve people on a jury, but with everyone—every American. So I want you to call the president, I want you to call your representatives in government, I want you to call the tobacco companies, the antitobacco lobby, whoever. And I want you to tell them that this is America and that you don't want to be told what you can and can't do in the privacy of your own home."

I began backing away, and the group went with me.

"Mr. Barnett, do you have any comment?"

"No. No, I really don't," I said, turning and starting for the car. I honked the horn as I threw myself into the driver's seat. "Anne! Come out and bring Nicotine with you! We're leaving!"

"I CAN'T BELIEVE IT, but it looks clear," Anne said, slipping out from behind a hedge and jumping back into the car. We were a few blocks from her apartment, and I'd sent her ahead for a little recon. The smart move would have been to go hide out in a hotel, but our current financial position prohibited such an extravagance.

I pulled away from the curb and rounded the corner, heading toward her parking space. "Thanks for putting me up."

"No problem. You can stay as long as you want. But this seems like an obvious hiding place. How long do you figure it'll take them to track you down?"

"Hopefully not until I've found a nice apartment in Outer Mongolia?"

"Bad food and worse weather. What's your backup plan?"

I parked the car and shut off the motor, but didn't get out. "That was my backup plan."

"Are you going to call Larry Mann?"

"What for? I don't work for Terra anymore. It's the worst of all worlds: I don't get paid and I don't have any power, but everyone thinks I'm responsible. Seriously, maybe we should take a trip. I've got some room on one of my credit cards. I always wanted to go to France . . ."

"The French smoke more than we do, and because of you we're not exporting."

"Yeah. Right." I climbed out of the car, and Nicotine clambered over the seats to follow. "Seems like it won't take long for the press to figure out I'm not in the loop anymore and move on. I'll just lie low for a few days."

"Mr. Barnett!"

We both turned toward the familiar voice and watched Blonde and Brunette jump from a car illegally parked next to a fire hydrant.

"Mr. Barnett! We need to talk to you," Blonde said as they jogged up to us.

"Geez, I haven't seen you guys since . . . When was it, Anne?"

"I guess it would have been when they ran away while we were getting shot at."

"Your father would like to talk to you," Brunette said, pretending not to hear.

"You know, my calendar's a little full right now. Tell him to have his people call my people, and we'll try to set something up for next month."

"It won't take long, Mr. Barnett. An hour at the most," Brunette said politely.

I started to turn away, and he grabbed my arm. "Really, sir. Just a short meeting. He's very anxious to get together with you."

I tried to jerk my arm back, but he just increased his grip.

"I said no! Now let go of me!"

"Why don't you just relax, Mr. Barnett?"

It was then I realized that these assholes were just a couple of steroid-enhanced pretty boys—neither of whom was even bigger than me. Fuck them.

Before Blonde could get ahold of my other arm, I pulled it back and rammed my fist into Brunette's evenly tanned face. He was so surprised, he didn't even try to duck.

My arm slipped from his grip and he fell back on the asphalt, blood flowing freely from the impressive piece of rhinoplasty centered on his face. Blonde made a move toward me, but then something happened that I'd never seen before. Nicotine jumped between us with teeth bared and hair bristling. A low growl escaped her and Blonde froze, staring down at one hundred and fifty pounds of muscle, fur, and teeth.

"Hello?" I heard Anne say. I glanced back and saw her talking into her cell phone. "Police? I'd like to report a kidnapping in progress."

Blonde wisely concluded that his momentum was lost and he backed away, helping his partner to his feet and then retreating with him to their car.

"Are you all right?" Anne said, hanging up the phone. She took my hand and looked down at the bloody knuckles. "Jesus, Trevor. What got into you?"

Nicotine continued to growl, and I used my other hand to smooth down the hair on her back.

"Do you have anything in your fridge?" I said. "I'm starving."

44 I SLEPT THAT NIGHT THE WAY I HAD WHEN I WAS A kid—a dreamless, careless preview of death. The loss of my job at the hands of my father had combined with my triumph over the Ponytail Brothers to more or less free me of the baggage I'd been carrying around since my childhood. While it was true that my future wasn't exactly so bright I had to wear shades, it was throwing off a warm glow. You know what I'd fallen asleep thinking about? I mean, other than Anne? Being a chef. No kidding. A chef. I'd never allowed myself to seriously consider it before. Change, I'd decided, was good.

A ringing phone wasn't enough to awake the new, temporarily-at-peace-with-himself Trevor Barnett, but Anne's elbow in my ribs was. I pulled my face out of the pillow and squinted at her. Dawn was beginning to burn through the window, and it provided enough light for me to see the top of her head poking out from beneath the covers.

"No one calls me this early," came her muffled voice through the quilt. "It's for you."

We were both too groggy to worry that someone had tracked me down, and I reached over her for the phone.

"Hello?"

"'Morning, Trevor! Beautiful day!"

It took me a moment to process the voice. Lawrence Mann.

"It's not day yet."

Anne rolled over and pulled her pillow over her head. Man, she was cute.

"You're not an easy guy to track down, Trevor. I've been calling your place but . . ."

"The press is still on my lawn. I guess I have you to thank for that." I propped myself into a sitting position against the wall. Anne's Smokeless Youth salary apparently precluded the purchase of a headboard. "Thanks a lot for the heads up."

"Jesus, Trevor, do you *ever* check your messages?"

"Not lately," I admitted. "It's almost always bad news."

"I couldn't tell you about this when we met because the final decision hadn't been made yet. But I did everything short of putting an APB out on you to try to talk to you before I made the announcement."

"Fine. But why, Larry? Why do this? You had your jobs back."

Anne pulled the pillow off her head and opened her eyes, listening to my side of the conversation.

"The hardest part of this thing was starting it—arresting my people's inertia and getting past their shortsightedness. You and Paul managed to force that. In the end, everybody was pretty resolved. The general consensus is that we've come this far . . ."

"Big risk, Larry. Are you sure it's worth it?"

"Sure, why not? If things start to fall apart, we can always go back to work. I've been thinking about this a lot, Trevor. My job isn't just in the here and now; it's in the future. In ten years, will there be a strong company that can offer my people a good place to work at a fair wage?"

"When all this started someone pointed out to me that Paul and I didn't have a friend in the world. Seems like that's your position now. My father, the board, the government, smokers—everyone wants this to go away. And here you are twisting a knife in them."

"Twisting it hard," Mann said. "I've set everything up to make it look like we're hunkering down for the long haul. When the press starts picking up on that, smokers are going to go ballistic. They thought they had their smokes back, and I pulled the carpet out from under them. I hear the

White House's switchboard has been completely lit up since I made the announcement and that lines to local politicians are fully jammed, too. Terra's headquarters has a hundred and fifty protestors out front."

"And what about your headquarters?"

He paused, and I imagined him walking over to his office window.

"About two fifty."

"Do they look angry?"

"You have no idea."

Anne threw an arm across me and wiggled a little closer.

"Well, good luck to you, Larry. I know you'll take it the right way when I say better you than me."

I leaned over to hang up the phone and had it mere inches from its cradle when I heard Larry's tinny voice.

"Hold on, Trevor! You started this. You can't abandon me now."

I frowned and put the phone back to my ear. "What?"

"I hear you might be in the market for a job."

I should have told him about my imminent enrollment in cooking school, but I didn't.

"We need a liaison between Terra management and the union. Are you available?"

45

I STILL HAD MY KEY TO TERRA'S EXECUTIVE ELEVA-
tor, and I figured I might as well use it. On the ride up from
the garage, I took a deep, cleansing breath and tried to forget the violent
shouts of the people outside. A few more breaths and I'd almost managed
to make my mind go blank. Any more thought about what lay ahead was
just going to make me more nervous.

When I stepped from the elevator, the two guards who had been so polite before charged me. I remained completely passive as they each
grabbed an arm and began forcing me back.

"I'm sorry, sir, but you're going to have to leave the premises immediately," one of them said.

"I have an appointment with Edwin Barnett," I said. "I'm representing
the Tobacco Workers' Union."

That seemed to confuse them.

While I had no hard evidence, I suspected that no one from the union
had bothered to mention to my father that I'd been hired and that I would
be coming in Larry Mann's place. Whether that was calculated to gain me
some kind of psychological advantage or just because Mann wanted to get
a dig in on my father, I wasn't sure.

"Feel free to call the union's headquarters if you want," I suggested.

They looked at each other for a moment and then just let me go. I
waited patiently as one of them punched in the code that opened the door

and then held it open for me. After I'd walked through, I swear I heard one of them say, "Go get 'em."

Probably just my imagination.

BASED ON MY father's secretary's invitation to "go right on in," I assumed that Dad was already aware of my presence in the building, but his reaction proved otherwise. He jumped up from behind his desk, an expression of fear briefly crossing his face, as though I was wrapped in dynamite and had the detonator under my thumb.

"What the hell are you doing here?"

"I signed on with Larry Mann yesterday," I said, falling into the chair in front of his desk. "He was busy, so he sent me in his place."

"This isn't a game, Trevor. You go back to Mann and tell him to get his ass in here. I'm not dealing with you on this."

I took one more deep breath and let it out slowly. On the drive in, I have to admit that I'd been more than a little excited about ramming some of the arrogance and insecurity my father had subjected me to right back down his throat. But now, actually sitting there, I just wanted the meeting to be over.

"Larry's not coming, and you can't afford to be picky about who you deal with. You're in trouble. Your takeover from Paul was clever and all, but it needed to go really smoothly in order to stick. I don't think I'd call this smooth, would you?"

"You're out of your league here, Trevor. I—"

"You tell me that every time I get knocked down, but I keep getting up, don't I? I'm starting to think you're the one out of your league. Are you watching the news? There's a lot of talk about the fact that this wasn't good for anyone but the lawyers and the politicians. Larry's been onto Senator Randal and the others, and they're scared shitless. You have money, but we have votes. Every politician in the South is about an inch from flip-flopping on this thing."

He looked confused, as if he couldn't decide whether to sit or to continue standing. "You have no idea what you're doing, Trevor. You're playing with the livelihoods of tens of thousands—maybe hundreds of thousands of people. Why? To get back at me? To try to get your trust back?"

I managed not to laugh, but it was kind of a struggle. "Your concern for the common man is almost as touching as your concern for the people out there dying of cancer, Dad."

"How long do you think these semieducated assholes are going to stand behind you, Trevor? There are no more cheap loans, no more partial salaries, no more subsidized food. No more stolen cigarettes. You put too much faith in them. They just want their Pabst Blue Ribbon and their pickup trucks. I can give them that. What can you give them? Security fifteen years down the road? These types of people don't think fifteen *minutes* down the road."

"I don't think they'll have to hold out for all that long. I can almost feel the board panicking. Can't you?"

"WHAT DO YOU want, Trevor? You want your job back? You want to be EVP-strategy again? You want your trust? Fine. You get these people back to work, and all that's yours."

"I don't think I'm going to be able to sell that, Dad."

"We can talk about some concessions to the union—give you something attractive to go back to Mann with."

I shook my head. "We're looking for a long-term solution."

"Bullshit!" he shouted. "This was all just an old man's fantasy! Congress and the president aren't going to override the courts and come out in favor of smoking, for Christ's sake!"

"They might be more receptive than you imagine," I said, sounding calmer than I felt. "I think we're in a position to make a deal that works for everyone."

"You can't do shit!" he said. "I run this company. Not you, not Mann, and sure as hell not a bunch of rednecks working in a factory! Do you hear me, Trevor? *I* run this company!"

46 I RAN THROUGH THE CRUSH OF PROTESTERS WITH MY hands on the shoulders of the enormous man in front of me. I was completely surrounded by Security and if I had had a towel over my head, I would have felt like a heavyweight champion headed for the ring.

Miraculously, we made it to the front door of the Tobacco Workers' Union headquarters and I ran through.

Conditions inside weren't much better than they were outside. People were everywhere: overflowing from cubicles, camped out on the floor, crammed into corners. Nearly everyone had a cell phone pressed to their ear, and all were speaking loudly and passionately into them.

"Welcome to Purgatory," Anne said, stepping over a woman who was kneeling on the floor pawing through a stack of loose papers.

I thumbed behind me at the door. "I think some of those people wanted to kill me!"

"I guarantee it," she said, motioning for me to follow her. She pointed to a man running cable through the ceiling as we made our way through the building. "All the lines to the published numbers are jammed."

I must have looked a little scared because she then added. "Hey, almost half of it's positive."

"What did the other half have to say?"

She didn't answer and since I didn't really want to know, I decided to change the subject. "Were you able to get in touch with Terra's board members?"

Not surprisingly, the showdown with my father hadn't gone well and I'd been forcibly ejected from the building after an unproductive meeting. Clearly, the thing that made sense for everyone was for Terra to team up with the union and continue what Paul Trainer had started. There was no way Dad was going to let that happen without a fight, though. He'd gotten his grip on his new office by opposing that strategy, and reversing himself now would make him kind of an irrelevant choice for CEO. He'd likely be demoted back to general counsel and if the industry won its battle for legal protection, he'd begin a long slide into obscurity.

"Yup. I talked to all of them," Anne said.

"Really? I figured most of them wouldn't take our calls."

"Nope. Talked to every one."

"What did they say? Did they agree to a meeting?"

She nodded.

"How soon? Can we get them to do it this week?"

"Oh, I think so."

"Do what you can, Anne. This thing's being held together with spit and bubble gum."

"How about today?"

"I admire your optimism, but I kinda doubt that's going to happen."

The board was made up of some of the wealthiest and most powerful people in the country. Somehow I didn't think they were all going to jump on their private jets and rush over here to meet with a thirty-two-year-old imposter.

"Buy me dinner if I can pull it off?"

"Sure."

"Where?"

It occurred to me that Larry hadn't brought up the issue of salary when he'd hired me. It seemed likely that I was working for free. "Might have to be McDonald's."

"You're on. They've been waiting for you in the conference room for an hour."

I stopped short, and a woman with an armload of posters took the opportunity to dart by.

"What?"

"They've been waiting for an hour. In the conference room."

A mild sense of panic I thought I'd become desensitized to suddenly made a forceful return. Anne must have seen it because she smiled and dropped the other shoe.

"Also, you've got Angus Scalia, John O'Byrne, and the heads of some of the other antitobacco groups waiting for you in Larry's office. They've been waiting for *two* hours. If you want my advice, you might want to talk to them first—I think they're about to kill Angus."

I considered my options. O'Byrne and his cohorts, who had doubtless spent the last two hours listening to an endless Angus Scalia diatribe on why they were wastes of skin, or the board who were men and women *very* unaccustomed to being made to wait—particularly by someone like me. Neither seemed all that attractive.

What about "none of the above"? Maybe it was time to start thinking outside the box a little. "Put the antitobacco lobby people in with the board."

"Excuse me?"

"Then, maybe you and I could go get some coffee." I walked off in search of the copy room. "Are there doughnuts?"

IT TURNS OUT they did have doughnuts—the little powdered-sugar ones that had been so instrumental in defining my current philosophy and raising my blood sugar. Anne sat on the counter, sipping coffee and chatting with the haggard people who came in to use the copier. I tried to do the same, but was having a hard time focusing. My master plan was to soften up the antitobacco people and the board by leaving them together for forty-five minutes. In the end, though, I could only sweat it out for half an hour.

"Do you think if you repeat that loud and often enough, somebody might believe it?" Scalia had his tiny feet up on the conference table, a precarious position for a man of his proportions. The other antitobacco people were on the same side of the table but were packed up at its edge, as far from him as possible. Sitting silently on the other side was Terra's board. I counted—they were all there. And they all looked pretty pissed off.

"I didn't come here to talk to you, and I have no idea what you're even doing here. So why don't you just—"

I'd done a little brush up and knew that the board member shouting at Scalia was Richard Peg, a former executive at Exxon. He fell silent when I entered. All eyes were on me.

"Ladies and gentlemen," I said smoothly. "For those of you I haven't met, my name is Trevor Barnett. I'm the former executive vice president of strategy at Terra, and now I'm working with Larry Mann here at the union." I managed to sound fairly calm. "I'm—"

"What is this crap?" Peg said, obviously still a little put out. "We all dropped what we were doing and came here with the promise of a meeting with you." He pointed across the table. "Not with them."

Scalia scowled. "Well maybe you should pucker up and—" I held up a hand and, surprisingly, he shut up.

"In fact, I had meetings scheduled with both you and the antitobacco lobby people you see here. I thought it might be more efficient to combine the two meetings."

"I don't see how having these people here is going to help us accomplish anything."

"Are you afraid we might hear your latest plan to double cigarette sales to grade-schoolers?" Scalia said, touching off a shouting match that included nearly everyone. I sat down at the head of the table.

"I'd like to answer Mr. Peg's question," I shouted, and the other voices in the room faded.

"As all of you know, the shutdown of the tobacco industry has caused a lot of grief over the last month. And that's only going to get worse as the shutdown continues." I pointed to myself and the board. "It's no secret what we want. We want ironclad protection from the constant lawsuits—"

The antitobacco people all started talking at once. O'Byrne, who had struck such a beneficial deal with my father, was particularly loud.

"Let me finish!"

Everyone fell silent again.

"The bottom line here is that we aren't going to get that protection for nothing, no matter how much we strong-arm the government. The purpose of this meeting is to hash out a deal that works for everyone. We want protection—and frankly we'll accept nothing less. You"—I pointed to the antitobacco lobbyists—"you want a level of concessions from us that will make that protection palatable. And that's fair. Now you talk. What, specifically, do you want?"

"I want you sued into oblivion," Scalia growled. No one responded to that statement, instead looking to me to do it.

"That's not on the table, Angus. And, frankly, I don't think it's good for anyone. I personally don't believe in putting a bunch of ambulance chasers in charge of setting policy in this country."

Yes, I'd finally done it—publicly expressed an actual, heartfelt belief without serious injury. Why not push my luck and try it again?

"I also don't believe in taking away Americans' right to do things that are bad for them. There's no telling where that would lead, but I suspect nowhere good. Look, the union is in this for the long haul. We either create a workable deal here or my people hunker down until the government cries uncle—sometime before elections, I'd guess."

Scalia didn't respond, but Peg did. "Then why don't we just hunker down and take our lumps, Trevor? Why offer concessions if we don't have to?"

"Because in the long term, we can't operate this way. We're in desperate need of a stable operating environment, and to get that we're going to have to create an honest and aboveboard culture."

He didn't argue, so I decided to open it up. "Okay, let's hear some productive discussion. Who wants to start?"

The silence stretched out for more than a minute, but I was determined not to fill it. Finally a woman from Terra's board spoke up.

"We could allocate, say, a billion dollars for research into producing a safer cigarette . . ."

That got a few of the antitobacco lobbyists nodding. Angus Scalia threw his legal pad at a wall and jumped to his feet. "That's the kind of utter shit that makes it impossible to deal with you people! How stupid do you think I am? You create 'lights' knowing that people will just smoke twice as much and you'll double your income while you watch them die just as fast. Now you want to be able to put out a 'safe cigarette' to prompt even *more* people to start smoking. And wait—let me guess—you want immunity from lawsuits when we find out ten years down the road that your 'safe' cigarettes weren't safe at all!"

"Angus!" I shouted. "Sit down!"

"I won't sit down!" He pointed at his antitobacco colleagues. "And you—you might be able to pull off some backroom deal to keep your jobs and your funding, but I'm going to make it my mission in life to expose you for what you are." He was forced to pause to catch his breath, and I used the opportunity to speak.

"I have to agree with Angus. This kind of thing isn't going to fly anymore. By putting legal protection in place, we're going to save billions of dollars in legal expenses, our stock price is going to go through the roof, and we're going to be able to operate in a predictable environment. In return, I think we can offer some real concessions that are going to have real consequences. And in the process, maybe shake off a twenty-five-year reputation of evil."

Scalia looked down at me, squinting behind his John Lennon glasses. Then he sat back down.

47

I'D SPENT THE LAST HALF HOUR WALKING AROUND my old floor at Terra's headquarters, peeking into cubicles trying to remember the faces of the people who'd worked in them. Everything looked as though it had been abandoned in a panic, though I knew that was a misinterpretation of the facts. A couple of days ago when my father took over, the employees had started to return, bringing with them their knickknacks, photographs, pillows, and the like. The suddenness and uncertainty of the strike had left the floor empty of people and full of partially unpacked boxes.

I still wasn't sure that I should be there—which was why I was hiding out on an empty floor. My new cadre of corporate spies had been giving me hourly reports of my father's desperate but ultimately doomed struggle to hold on to power. It had become kind of depressing, and I wanted to put a stop to it—to convince him that it was in his best interests to join me and Larry in continuing Paul Trainer's grand social experiment. Things were getting dangerous for everyone, and there was safety in numbers. United we stand . . .

I DIDN'T HAVE an appointment, but talking my way past the executive-floor guards turned out to be easy. A little too easy, actually. I felt myself slowing as I padded across the thick carpet, wanting to put off the coming confrontation as long as I could.

"Trevor! How you doin', boy?"

I stopped short, but was reluctant to acknowledge the familiar voice.

"What's the matter? Cat got your tongue?"

"What are you doing here, Paul?"

Trainer was sitting in an empty cubicle wearing a grin so wide that it seemed to extend past the edges of his face.

"Goddamn, son! I knew you wouldn't let me down. Brilliant! You had it in you the whole time—I saw it right off. You really are worth your weight in gold."

I should have seen this coming, and honestly on some level I probably had. With the strike in place, the board found their control over the company's destiny tenuous at best. Add to that the concessions I planned to make, and Trainer's return had been pretty much inevitable.

"I hear you got all kinds of ideas these days, Trevor." He got up and walked over to me, slapping me on the back. "You really are a kick in the ass! Now, come on."

"Where are we going?"

"To take our place in history."

My FATHER was on the phone when I walked in. He looked surprised and gave me a glare that didn't bite anymore, but didn't hang up. When Paul came in behind me, though, the handset hit the cradle pretty fast.

"What the hell—"

"You're sitting in my chair," Trainer said. He didn't sound angry, but his voice did have a coldness to it that I hadn't heard before.

"*Your* chair?"

"You screwed up bad, Edwin. What made you think I'd just roll over and let you take my company from me?"

Probably the fact that it was exactly what he'd done. If I remembered correctly, Larry Mann and I were the people who'd stood up and fought.

Mann laughed. "You're turning into a real tough guy, you know that? What happened to the confused little boy I met a few weeks ago? I'm serious, Trevor. You're the only guy in this whole goddamn industry who even understands the concept of a straight answer. Give."

"Turns out Paul Trainer's nuts," I said.

"And this is news? What I want to know is how all this fell apart. You'd met with the president, you had Congress worried about their jobs, the press and the country were coming over to your side. Why give up now?"

"My father is afraid of becoming irrelevant," I said. "And the board's scared."

"I see . . . Well, what about us, Trevor? What happens to us?"

"You got what you wanted. You got your jobs back."

"But for how long? We're going to lose in Montana. How long until the industry gets sued into oblivion? How long before Terra can't raise capital? Can't get loans? When we start to get squeezed, it isn't going to be management that feels it; it's going to be my people."

I shrugged at him in the rearview mirror. "I feel for you, Larry, but none of this is my problem anymore."

43 "WHAT'S UP WITH THIS?" I SAID, BANGING ON THE steering wheel in frustration. Instead of dwindling, my media contingent had doubled in size, and the appearance of my car had reporters jumping from their vans and scurrying around as though someone had thrown a grenade.

"Look at my lawn, Anne! How am I gonna sell this place now that they've destroyed my damn lawn?"

"Maybe you could put in one of those Japanese rock gardens."

I slammed on the brakes, bringing the car to a lurching halt a hundred yards from my driveway. "What the hell's wrong with them? I'm fired and cigarettes are going back on sale tomorrow. I don't have anything else to say." I shoved the car into reverse and was about to do an impression of my former bodyguards when Anne pulled the shifter into neutral.

"Hey!" I said over the sound of the revving engine. "What are you doing?"

"Maybe you should talk to them."

"About what? I'm not the company's spokesman anymore. I'm just a tobacco industry heckler, right? And not even a good one."

"You've got potential," she said through a broad smile. "It just takes a little practice."

Through the windshield I confirmed that the reporters weren't advancing on us. Yet.

"Have you ever thought of just speaking for yourself, Trevor? It might be a good time for you to get on TV and sound magnanimous and intelligent and unflappable. To tell people what you were trying to do."

"What's the point?" I said.

"First of all, I'll bet it will make you feel better. And even if it doesn't, you never know when a future employer might be watching. It seems to me that the combined income of the people in this car is zero. And I like to eat. Remember?"

Combined. I liked the sound of that.

I put the car in drive again and glided regally forward, trying to get in a magnanimous, intelligent, and unflappable frame of mind. It turned out to be kind of a stretch.

The reporters crowded the car, shouting and attacking the windows with their microphones. Instead of pulling through them into the garage, like I normally did, I forced my way out onto the driveway. I motioned for quiet and, for some reason, it worked.

"Okay," I said loudly. "I'm going to stand here and answer every question you've got as honestly and completely as I can. But after that you've got to leave and let my lawn heal, okay?"

Very magnanimous.

"Mr. Barnett—" one of them started, but I held up my hand for silence again. "Let me just say a few things before we start the Q and A. First, I am no longer employed by Terracorp and so I no longer speak for them."

"Were you fired?"

I thought about that for a moment but couldn't see a way to sugar coat it. If anyone in history had ever been fired, it was me. "Yeah, I think it would be fair to say that I was fired—"

"By your own father? How did that make you feel?"

"Unemployed. It made me feel very unemployed. Look, the strategy that Paul Trainer and I came up with"—I figured I might as well take some credit since I was in the market for a new job—"turned out not to be viable in the board's eyes and they've decided to go back to the company's former policies. Frankly, I don't agree with that decision. I thought we had a real opportunity for everyone in this country to get on the same page where the risks of smoking are concerned. Unfortunately, when push came to shove, the powers that be didn't share my vision."

"Do you feel that this is a step backward for the interests of the tobacco companies?" I recognized the guy who'd asked that question. A financial reporter—I couldn't remember from where. It was hard not to think about the fact that what I said next would either make or lose millions of dollars for tobacco industry shareholders—a group I was no longer a member of.

"Honestly, I don't think it's in the best interests of anyone, but it's the board's call and I have to respect their decision."

"Would you elaborate on that, Mr. Barnett?"

The competition in the group seemed to be easing as it became clear that I would keep my word and answer all their questions. The increased calm allowed me to relax a bit.

"People know that smoking is bad for them—if they can't tell by the way it makes them feel, then they can read the big warning on the pack. What we've lost here is an opportunity to create an environment of personal responsibility that perhaps could have reduced the smoking rates in this country, stabilized an economically critical American industry, and stopped wasting the courts' time in endless lawsuits."

"Or an opportunity to completely do away with the tobacco industry," someone said. I saw Anne slip out of the car and make a break for the house. No one else noticed.

"Maybe," I said. "That was for the American people and their representatives to decide."

I felt like I could now check off "unflappable."

"But isn't it true that your real goal was to create an environment that would allow the tobacco companies to sell a dangerous product without fear of litigation?"

"In a word—yes. It's true that we make a dangerous product. If you smoke, there's a pretty good chance it's going to kill you one day. And if

At the first hint of personal danger, Trainer had cut loose everything and saved himself.

"This company is on the verge of bankruptcy," my father protested. "You were driving it under! There's no way the government is going to give us full protection. I didn't take this company from you. I saved it from you."

"Took it. Saved it. Doesn't really matter anymore. The board's reinstated me. You're out."

"What? I—"

"Don't talk, Edwin! Just get the hell out of my office. You're fired!"

I could see my father's knuckles turn white as he squeezed the edges of the desk.

"You can't fire me. My family started—"

"You've got ten seconds to get your ass out of my sight before I call Security."

Trainer had the good manners not to count out loud, but he seemed to mean what he said. My dad stood his ground for three or four more seconds and then came out from behind the desk and headed for the door. He didn't look at me as he passed.

"DAMN, THAT FELT GOOD!" Trainer said, sitting down in his chair and sending my father's things cascading onto the carpet with a sweep of his hand. "And what about you, Trevor? This is a red-letter day, huh? If I remember the way your trust is written, you just got all your father's money *and* his house."

"I don't want my father fired," I heard myself say.

"What?"

"I don't want him fired, Paul. He's still my father and I have my mom to think about. I don't want him fired."

Trainer screwed his old face up, clearly not understanding my willingness to walk away from a perfectly good piece of revenge.

"Do I look like I work in Personnel?" he said finally. "You're an executive vice president. You want to rehire him, then rehire him. But not in Legal. If I ever lay eyes on that worthless son of a bitch again, he's gone permanently. You understand?"

I nodded.

"Now what do you say we call the White House and set us up a meeting? It's time to finish this thing."

WHEN I WALKED out of Trainer's office a few minutes later, there had still been no acknowledgment that a couple of days ago he'd completely screwed me without a second thought. And while there had been a time when his charisma and the sheer intensity with which he ignored that simple historical fact would have confused me, that time was long past.

I found my father in his office yelling into the phone.

"I don't give a shit if he's in a meeting! Tell him that Edwin Barnett is on the phone and it's urgent!"

He looked up at me and I leaned against his doorjamb, unwilling to enter.

"What do you want?"

"Hang up the phone, Dad."

"What?"

"Hang up the goddamn phone!"

He jerked fully upright and, after a few seconds of pointless defiance, did as I said.

"Jesus Christ, Trevor. No one will take my calls. I'll lose everything . . . Are you proud of yourself? What about your mother? She'll lose everything, too. Don't forget that. If I go down, she goes—"

"I'm rehiring you, Dad," I said before he could start begging. That was an experience I could definitely live without.

"What?"

"You're rehired, okay? I'll have our lawyers draw up a contract that'll put you on the books as an employee until your trust is distributed. But you've got to get out of here now and never come back. If I ever hear anything about you trying to contact a board member again, I'll fire you and take everything. And if Paul Trainer ever so much as lays eyes on you again, he's assured me that he'll fire you on the spot. Do you understand?"

"Trevor, I—"

"Do you understand?"

"Yes."

"TRAINER'S BACK," I said quietly into the phone.

"I guess we all knew that was coming," Anne said. "I suppose he heard about your meeting with the board and the antitobacco people. What happened to your dad?"

"Trainer fired him and I rehired him. He'll get to hold on to his trust, but he won't have anything to do with the company anymore."

"That's good, Trevor. That's the right thing to do. He is your father."

"I guess."

"So what are you going to do now that Trainer's back in charge?"

"I don't know."

"Some good things came out of that meeting yesterday, Trevor. Some things that could work. Don't give up now."

"Don't worry, I'm through giving up. Is Larry there?"

"Yeah, you want to talk to him?"

"Please."

The line went silent for a few seconds, and then Larry picked up.

"Trevor! I hear Trainer's crawled back out from under his rock."

"Yup."

"How's that going to work out?"

"I suppose it's up to you."

"Me? How you figure?"

"Who are you backing, Larry? Me or Trainer?"

Mann laughed into the phone. "Now you're getting it, Trevor. Good for you. I'll tell you this: Every time I shake hands with Paul Trainer, I count my fingers afterward. You're my horse, Trevor. You are."

48 THIS TIME OUR ESCORT DIDN'T SEEM AS PRETTY, AND I could see the wear patterns in the carpet. It was a Saturday and the White House was quiet, providing me an opportunity to try to figure out why I was there. I'd hatched ten different plans to get myself included in this meeting but, in the end, hadn't needed to use any of them.

It had seemed reasonable to assume that Trainer would try to freeze me out at the last minute so he could present his rather one-sided agenda without complications. Maybe he thought I'd just sit there with my mouth hanging open like I had last time. But still, why take the chance? He didn't need me.

President Anderson, dressed in khaki slacks and a golf shirt, was standing in the middle of his office when we walked in. Trainer had opted for a black suit and resembled the mortician from a horror movie I'd once seen. *Pet Semetary*? No. *Phantasm.*

"How are you, Paul?" Anderson said, taking Trainer's hand. "I see you managed to come out on top again. Why did I never doubt it?"

"It's good to see you, Mr. President." Trainer motioned toward me with his head. "You remember Trevor Barnett, don't you?"

"Of course."

Anderson's grip was firmer and longer than I remembered.

"I'm so glad you could make it."

His tone and inflection answered the question I'd been contemplating: He'd requested that I be there. But why?

"Well, Paul," Anderson said, taking a seat on a sofa in the middle of the room and motioning for us to sit on the one across from him. "You seem to have managed to force the showdown everyone wanted to avoid. I suppose congratulations are in order?"

"I guess not everyone wanted to avoid it, Mr. President."

Anderson nodded. "I'll ask you again to put your people back to work while we try to hash this thing out. Call it a cooling-off period."

"I don't think I can, sir. This strike was called with absolutely no involvement from me. I have no control over Mann and the union."

To my surprise, Anderson turned toward me. "What do you think, Trevor? Can we shelve this strike for a little while?"

"Uh . . . I don't think so, Mr. President. Larry's aware that once everyone's back to work, it'll be really hard to get them to walk off again—and even harder to reestablish support from the Teamsters. The Tobacco Workers are looking for a solution that'll preserve their way of life going forward, and they're willing to fight hard for it."

"You bet they will," Trainer said, but Anderson ignored him and stayed focused on me. It seemed that he understood what I hadn't managed to fully grasp. Trainer wasn't the power in this room. I was.

"Come on, Trevor. You have Larry Mann's ear—we all know that. What can I do for you that would persuade you to talk him into making this thing go away for a little while? He'll listen to you."

I shook my head. "The reason I have the trust of labor is because I've been straight with them. If I stray from that, Larry'll see right through me and it'll get us nowhere. I just don't think we're going to be able to make this disappear."

"No, I suppose not," Anderson said. "Larry Mann's a hard person to influence."

I'd discovered that the government had subtly threatened our board members with things like audits, in-depth examinations of their legal and business dealings, and even scrutiny into the activities of some of their less-than-angelic children in order to get them to oust Trainer and get cigarettes flowing again. I'm sure they'd tried the same thing with Larry and found, like Terra had, that he was squeaky clean and truly concerned with the welfare of his people.

Anderson leaned back in the sofa's cushions and crossed his legs. "Okay. What should we be looking to accomplish in this meeting?"

"I think we should be looking to work out a deal, Mr. President."

"Damn right," Trainer said, a little too loudly.

"And that deal," Anderson said, "starts with ironclad immunity from lawsuits relating to smoking."

Trainer nodded.

"Okay, let's assume for a moment that's possible. What are you going to give me in return?"

Paul shot me a cautioning glance before he spoke. "A billion dollars over the next ten years in safer-cigarette research and a significant increase in the industry's funding of the battle against teen smoking, as well as support for stronger legal penalties for underage smokers."

"And I assume there will be a thousand pages of legal strings attached to those, um, compromises?"

"Well, obviously it needs to be a deal that makes sense . . ."

"Okay. What else?"

"What else?" Trainer said, feigning shock that anyone would even ask such an absurd question. "Mr. President, those are enormous concessions that have the potential to significantly impact our business . . ."

Anderson laughed at that. "Who do you think you're talking to, Paul?"

Trainer shrugged his skinny shoulders. "I'm not sure what else to say. I'd be irresponsible to give up more. Maybe we should pull back and give this thing a little more thought—get together again in a few months."

A bald-faced threat. A few months would take us right up to the elections.

Anderson didn't immediately react. After what seemed like a long time, he turned back to me. "What do you think, Trevor? Should I take that deal?"

I could feel Trainer's eyes burning into the side of my face.

"No, actually, I think that's a pretty bad bargain for everyone."

"I'm sorry, Mr. President, can I have a word with Trevor outside?"

"In a minute," Anderson said.

Trainer fidgeted like a child. "You understand that Trevor is someone I just brought in recently. He has no real—"

"Paul," Anderson cautioned, "Trevor has the floor."

I cleared my throat. "If you take that deal, I can guarantee you that Angus Scalia and the rest of the antitobacco lobby will come out hard against the safer-cigarette research—paint it as a ploy to convince people to start smoking and smoke more. They also won't accept any funding for the fight against teen smoking if it's tied to company profits—and Paul will insist on that."

Trainer snorted. "They'll accept anything we give them."

"Not this time, Paul."

"So what do you propose, Trevor?"

"He can't propose anything," Trainer said. "He has no power to speak for this industry other than the power I give him."

I shrugged. "I guess I take the opposite view. Who do you speak for, Paul? The board was happy to sacrifice you, and you have no relationship at all with labor. You're only back in your office because of me."

His jaw dropped. And I mean this literally. He was sitting there, five feet from the president of the United States, with his mouth hanging open. What he undoubtedly wanted to do was tell me that I was just some kid he'd used to dupe the public and put in the line of any potential gunfire, but he knew it probably wouldn't play very well in front of Anderson.

"Mr. President, I have the complete support of the board now. No one—and I mean no one—thinks that Trevor Barnett is capable of running Terra or taking a leadership role in the industry."

I mimicked Anderson by crossing my legs and leaning back into the sofa cushions. "The board brought Paul here back because they're afraid I'll give away too much. A problem for them but a benefit for you, Mr. President."

"I'd have to agree," Anderson said. "Give me some specifics, Trevor."

"Mr. President—"

"Be quiet, Paul! Go ahead, Trevor."

"We'll accept combined state and federal taxes that will bring pack prices up to five dollars nationally. That'll create a real deterrent to smoking—particularly by young people. It'll also create billions of dollars in additional tax income."

Anderson frowned. "I can't dictate to the states how much they tax you. Hell, it's seven fifty a pack in New York right now."

"New York will have to back off. But I'm guessing they'll not want to be the spoilers on this thing when they've got a number of guys up for re-election."

"Okay," Anderson said. "Go on."

"We'll accept that the protection we get from suits will only be for people who started smoking after the government warning labels were put on the packs. Anyone who started before that can still sue us, but that right won't accrue to the relatives of people who've died from smoking-related illnesses. We think that a fair compromise. Clearly people did get addicted while we were playing down the dangers, but if we allow relatives to sue for pain and suffering we'll end up fighting thousands of suits by people who say they're depressed because one of their ancestors died from pipe smoking during the Civil War."

Anderson nodded but didn't say anything.

"And we'll need a cap on annual punitive damages. They're getting out of hand."

"How much?"

"I'll have to talk to my people, but the number's going to be really low. We're already paying out more than we made from 1950 to present to the states. We've been punished, and it's time to move on."

He nodded again, prompting me to continue.

"We'll agree to a nationwide ban on all print advertising—"

Not surprisingly, that was more than Trainer could take. He grabbed my arm and dragged me to my feet. Anderson didn't say anything, so I allowed myself to be pulled over to a far wall.

"What the fuck are you doing, boy?" he said in a harsh, smoker's whisper. "Still trying to buff off that soul of yours?"

"What I'm doing, Paul, is making this a company that's going to be viable in the twenty-first century, not the eighteenth. Print ads account for six percent of our marketing budget and about ninety percent of our bad press. We need to clean the slate here and start over."

"You don't have any fucking idea what we need to do! For the rest of this meeting you're just going to sit there and shut up. Do you understand me?"

I yanked my arm free. "Why don't you go back to cowering in that greenhouse of yours, Paul? Because if we go head-to-head on this and you lose, I'll make it my life's ambition to see that you spend what's left of your life driving around in a beat-up RV trying to keep ahead of your ex-wives."

I couldn't believe I'd just threatened Paul Trainer. Apparently, he couldn't either. Unsure what to do, he turned back toward Anderson, who had been watching us from his sofa.

"Mr. President—"

"I'll tell you what, Paul," the president said. "Why don't you excuse us for a little while."

Like magic, a Secret Service agent came through the door and escorted Paul Trainer out into the waiting area.

Anderson pointed to the sofa across from him again, and I sat.

"You were saying something about a ban on print media?"

"Uh-yeah. I mean, yes sir. We'll adhere to a voluntary ban, which I think you'll agree is a huge concession . . ."

"I do agree. But in my experience, huge concessions always come with huge strings attached. What do you want in return?"

"Ironclad protection against secondhand-smoke suits, too."

A slight frown.

"Sir, as hard as they've tried, neither the World Health Organization or the American Cancer Society has been able to tie environmental tobacco smoke to health issues. Without this protection, we've gained nothing. The lawyers will just shift their focus."

His frown didn't deepen but neither did it disappear.

"We're also willing to make a one-time donation of three hundred million dollars spread out among various antismoking groups," I continued. "There will be no strings attached to this at all—it's not a series of payments based on our profitability, or anything else. They just get the money."

He nodded. "Is that it?"

I tried to achieve an easy smile, but I don't think it worked. "We want teen smoking decriminalized."

He blinked a few times "What?"

This one was for Anne. I guess she'd been stung more than I thought when I suggested that the antismoking lobby's real purpose was more to stand on the moral high ground than to reducing smoking. She'd decided to take a few political lumps in an effort to make a difference.

"Every study shows that criminalization makes teens smoke more," I said. "The higher the fines, the higher the smoking rates. The antismoking groups all agree that it's time to decriminalize teen smoking and work with kids on a grassroots level. They think they can make real inroads here."

Anderson appeared to be in pain. "I know everything you're saying is true, Trevor, but what's the point? It just muddles the issue."

"But it offers a real chance for reducing underage smoking."

"It's too complicated to get across in a sound bite. People won't understand."

"Angus Scalia is heavily behind this provision, Mr. President, and the press loves him. Combine that with an industry-supported initiative to get the smoking age raised to nineteen to keep tobacco out of the high schools, and I think we can get this done."

Anderson sighed quietly. "Fine. But this is your responsibility, Trevor. I'm telling you up-front that I'll back away from this in a second if there's any backlash."

"I understand, sir."

His expression turned thoughtful. "You're giving away a lot, Trevor—more than you need to, maybe. Are you sure you can sell this to your people?"

I nodded. "We're giving up a lot, but we're also getting a lot. Some of those concessions might eventually translate into a decline in sales, but it'll be a slow, steady decline. A certain future, even if it's not an ideal one, is pretty attractive right now."

Anderson was silent for a little while. Finally he stood and offered his hand. "Okay, Trevor. You've sold me. You put your people back to work, and I'll throw my full weight behind this."

I have to admit, he was good. He'd spoken with such casual authority that I almost agreed without thinking.

"I'm afraid not, sir. When this is signed into law, then we'll go back to work."

49 PAUL TRAINER HAD BEEN LEFT IN THE OVAL OFfice's waiting room for more than half an hour, only to be called in to shake the president's hand and be dismissed. He stalked along behind our escort, clenching and unclenching his fists, not saying a word. I guessed that was because he was busy formulating an elaborate plan to have me killed.

We were deposited unceremoniously in the parking area, and Trainer waited until we were alone before he let loose.

"If you think you're going to take this company away from me, you've got another thing coming, boy. Don't believe the legend I've given you—I picked you because you'd bend over and do anything I told you to. And because you were stupid enough to look sincere doing it."

I was glad he'd finally been able to get that off his chest. A few more minutes and I think he might have stroked out.

"I'm going to *destroy* you," he continued. "Do you understand me? And I don't just mean you're going to lose your job and your trust—I mean I'm going to see to it that you lose everything! You'll never get a job paying more than fifteen grand a year as long as you live. I promise you that."

My brow knitted a bit as I considered his words. Not the threats—I didn't really care about them—but the thing about me taking the company away from him. Honestly, until he mentioned it, I'd never considered that I might end up running Terra. I was completely unqualified and, frankly, completely unmotivated. In order to hold this deal together, though, it was possible that at the ripe old age of thirty-two, I might have to insert myself as the CEO of one of the world's largest corporations.

I must have smiled at the thought, because Trainer blew yet another a gasket.

"Who the fuck do you think you are?" he shouted, ignoring the fact that we were almost certainly being watched. "You think you can take me on? Is that what you think?"

"Retire, Paul. Go play some golf and live the good life. You're entitled."

"This isn't over, boy. This ain't anywhere near over. You don't have a ch—"

"Wake *up*, Paul! I've got the union in my pocket and every politician in the South who has any kind of instinct for self-preservation is going to follow those votes right off a cliff if I tell them to. The antitobacco lobby is going to back me, the president is going to back me . . ."

Trainer opened his mouth to speak, but I cut him off. "Have I forgotten anyone? Oh, right, our shareholders. With a long-term solution to our legal problems, I can pretty much guarantee our stock prices are headed straight up. So they love me, too. Who's that leave you with? Some militant smoker living in a cabin in Idaho?"

"The board won't—"

"Jesus Christ, Paul—I just got the government off their backs and everyone in America on their side. What do you have to offer? Nothing."

"You think so, huh? Don't get too comfortable, boy."

Trainer suddenly burst forward and leaped into the back of the limousine, shouting "Go! Go!" before slamming the door behind him. The driver shot me a frightened glance through his open window but didn't move.

It was sort of sad. Trainer yelling "Go! Go!" and the limo just sitting there with the engine idling.

I waved the driver on and watched the car glide away without me. Trainer was undoubtedly already dialing his cell phone and formulating a plan that would allow him to claw his way back into power. By leaving me standing in the middle of D.C. without a clear way to get home, he figured he'd delay my efforts to counter him.

I sighed quietly and took the footpath through the gates and out to the road. I found a bench on a corner a few blocks away and sat there watching the cars and people go by.

About half an hour had passed when a large white limousine glided to a stop next to the curb in front of me. The back window descended smoothly, and Anne stuck her head out. "Need a ride, sailor?"

"Nice car," I said. "I was expecting a beat-up Ford Fiesta."

She threw the door open. "I figured it's company money, so why not travel in style?"

"That's what you figured, huh." I slid in next to her and closed the door. The glass separating us from the front was heavily smoked, and I couldn't see even a hint of the driver as we merged into traffic.

"So," Anne said, pushing me back in the seat and straddling me, "how's the prez?"

"He's okay." I was finding it a little hard to concentrate on our conversation in this position. "He wasn't happy about the teen-smoking thing, but it was hard for him to argue since I'd pretty much given away the farm at that point."

"I knew you wouldn't let me down."

"Did you get the welfare system going again?"

I'd put her in charge of getting Trainer's cheap food and loan program back on line. There was just no way you could be too popular with the union.

"All done."

"And the cigarettes?"

"I put Larry back in charge of security for Terra's warehouses. I'm guessing they're pilfering smokes by the truckload as we speak."

I slid my arms around her and pulled her close. Our lips were nearly touching now, and I glanced up to confirm that the driver couldn't see us. "Was there anything else? I forget . . ."

"Just your press conference."

That would be the one where I was going to blather on about how much the government was looking out for people's interests and wax rhapsodic about the new era of honesty and philanthropy that I was going to usher in. After that, Scalia and the other antitobacco pundits, as well as a few handpicked Wall Street analysts, would hit the talk-show circuit and effusively support the deal. I'd already brought back our publicity and marketing people and they were working 'round the clock.

"When is it?"

"Four o'clock."

"Fine, whatever." I leaned forward and tried to kiss her but she pulled back, smiling mischievously. "We dumped the five million kids smoking campaign. You want to see what we're going to replace it with?"

"I could wait, actually."

"No you can't."

She grabbed a rolled-up poster from the floor and pulled the rubber band off it with all the drama of a striptease. I rose up on my elbows and examined the reasonably, but not spectacularly, pretty woman centered on it. She was staring at a lit cigarette in her hand with a mix of suspicion and fear, holding it slightly away from her as though it might attack at any minute. Across the bottom, in big, bold lettering, the slogan read: YOU'RE ON YOUR OWN, BABY.

EPILOGUE

TWO YEARS LATER AS I WRITE THIS, THE MEMORY OF EVERY-
thing that happened is much clearer than the memory of who I used to be.
So much has changed.

The bill protecting the tobacco industry was signed into law an amazing forty-three days after I met with the president. And because it was
partially a clever clarification of the existing laws on the subject (thanks to
my good friend Dan Alexander), it had a somewhat retroactive effect.
The result was that a good ninety percent of the suits against the industry
were dropped as the attorneys bringing them saw the chances of recoup-
ing their outlay sink to almost zero. Don't worry about them, though—
they landed on their feet and already have the fast-food, candy, and
soft-drink industries quaking in their boots.

The infamous Montana suit wasn't dropped and we lost, but the judge
bowed to political pressure and reduced the award to an amount that the
industry could bond off. And while our appeal isn't yet completed, it
seems almost certain that we'll win. Actually, the plaintiffs' attorneys
made a quiet offer to settle, but we decided to go ahead and try to kill it in
the courts.

Anne is now the co-director of Smokeless Youth and the driving force
behind their highly visible Team Teen project. The stiff legal penalties
aimed at kids are all gone now, and Anne is working in an advisory capacity to Team Teen, the actual head of which is a very bright and dedi-

cated seventeen-year-old named Cindi who has struck fear into the heart of Big Tobacco like no one before her. There are no real numbers yet but increased prices, combined with an ad campaign depicting kids buying smokes from dimwitted adult store clerks and the slogan "You're smart enough to get them, are you smart enough not to?" seems to be striking a chord with young people all over America.

Paul Trainer put up a fierce, but ultimately pointless, fight to retain the helm of Terra. To his credit, it took him only a few days to realize that his time was over and that the few people who bothered to take his calls at all were just patronizing him. Did I make good on my threat to leave him fleeing his ex-wives in a '72 VW van? Nah. He's living in a ten-thousand-square-foot house in New Mexico. I understand that he's shooting in the low nineties and that he cheats.

My father is doing roughly the same, though I understand he cheats even worse. The employment contract I had drawn up allowed him to keep his trust, which will soon be fully distributed to him. I haven't seen or spoken to him in more than a year and a half. This is also true of Darius, though not because I'm still mad at him. I've just outgrown him.

Larry Mann is still the head of the Tobacco Workers' Union and still a champion of the changes we put in place. Terra and the other companies have stabilized and stock prices have skyrocketed, which is good for everyone, I think.

What about me? In the end, I actually did take the job as CEO of Terra for a short time. The first month and a half of my tenure were consumed with lobbying, and the next three were spent instituting the concessions I'd promised fully and fairly. The rest of my time at Terra was spent finding a replacement. Believe it or not, I managed to convince the CEO of Ben & Jerry's Ice Cream—a brilliant guy and die-hard Grateful Dead fan—to take the job. He is now in the hilarious and surprisingly effective process of turning Terra into the employee- and environment-friendly corporate citizen its name suggests it is. If he isn't careful, he's going to make Big Tobacco so popular people will start smoking just to support it.

Not surprisingly, I jumped ship the minute the new CEO was settled in—though I have to admit to my fall was broken by a golden parachute so large I hesitate to describe it here. Let's just say that I now have the ability to write much larger checks than I could before and I no longer feel the need to sign them in red.

Shortly after my departure, and based solely on the strength of my chili pepper and wild mushroom soufflé, I was accepted to France's most prestigious culinary school. My graduation four months ago was one of the proudest days of my life.

Me and Annie? Our relationship turned out to be one of those rare things that's even better in reality than in fantasy. No, we aren't married yet, but that's the result of a lack of time rather than a lack of commitment. I tried to convince her to quit her job and come to Paris with me but she refused, citing all the lives she was saving and offering to keep Nicotine for me. Still such a believer.

We moved in together when I returned to the States, though between her travel for SY and me starting a business, we haven't seen much of each other. The doors of my restaurant—which reviewers are calling one of the best in the Carolinas—finally opened a few weeks ago, and I'm already pretty certain that it has no hope of ever turning a profit. I don't really care, though—I've got more money than I could spend in two lifetimes.

Oh, and at Anne's insistence, I finally gave up smoking for good. It turned out that it was easy.

ABOUT THE AUTHOR

Kyle Mills lives in Jackson Hole, Wyoming, where he spends time skiing, rock climbing, and writing. He is the best-selling author of *Sphere of Influence*, *Rising Phoenix*, *Storming Heaven*, *Free Fall*, and *Burn Factor*.